Dude, Can You Count?

Dude, Can You Count?
Stories, Challenges, and Adventures in Mathematics
CHRISTIAN CONSTANDA

C

COPERNICUS BOOKS

An Imprint of Springer Science+Business Media

Christian Constanda
The Charles W. Oliphant Endowed Chair
  in Mathematical Sciences
The University of Tulsa
800 South Tucker Drive
Tulsa, Oklahoma 74104
USA
christian-constanda@utulsa.edu

Disclaimer:

All characters in this book are fictitious. Any resemblance to real persons, living, dead, or otherwise, is purely coincidental and unintentional. The book expresses the opinions of the author and is not intended to malign any religion, ethnic group, organization, or individual in this or any other universe.

Published in The Netherlands by Copernicus Books,
An imprint of Springer Science+Business Media, LLC

ISBN 978-1-84882-538-3   e-ISBN 978-1-84882-539-0
Springer London Dordrecht Heidelberg New York

British Library Cataloguing in Publication Data
A catalogue record for this book is available from the British Library

Library of Congress Control Number: 2009939689

Mathematics Subject Classification (2010): 97A20, 97B20

© Springer-Verlag London Limited 2009
Apart from any fair dealing for the purposes of research or private study, or criticism or review, as permitted under the Copyright, Designs and Patents Act 1988, this publication may only be reproduced, stored or transmitted, in any form or by any means, with the prior permission in writing of the publishers, or in the case of reprographic reproduction in accordance with the terms of licenses issued by the Copyright Licensing Agency. Enquiries concerning reproduction outside those terms should be sent to the publishers.
The use of registered names, trademarks, etc., in this publication does not imply, even in the absence of a specific statement, that such names are exempt from the relevant laws and regulations and therefore free for general use.
The publisher makes no representation, express or implied, with regard to the accuracy of the information contained in this book and cannot accept any legal responsibility or liability for any errors or omissions that may be made.

Printed on acid-free paper

Springer is part of Springer Science+Business Media (www.springer.com)

For Lia and Dan

Preface

> *All that is necessary for the triumph of evil is that good men do nothing.*
> Edmund Burke[1]
>
> *Those who think they know everything are a great nuisance to those of us who do.*
> Isaac Asimov[2]

Dear Reader,

There is no doubt: the world is going to the dogs.

Modern civilization, as we know it, is being steadily devoured by the hungry demons of sloth and apathy. What we inherited from the ancient Greeks and then the Romans, the proud tradition of arts and sciences, of culture and law, commands less and less respect as time goes by, with fewer and fewer fellow citizens willing to preserve it and arrest its passing into oblivion. Unconvinced? Just take a look at our institutions and judge for yourself.

People monitoring the general malaise of the human species might regard the current trends in public school education as one of its symptoms. Personally, I don't believe so. I think they are one of its causes. At least where mathematics is concerned, my many friends and colleagues around the world and I see evidence of their damaging effects every year: courses need to be watered down, remedial classes become a necessity, examination standards are lowered in line with performance expectations, grades are inflated to hide the sad truth, and so on. Those of us who do not subscribe to such unprincipled strategies earn nicknames like Darth Vader, Attila the Hun, Voldemort, or The Terminator. If this is the price we must pay for trying to make society understand that without adequate knowledge of mathematics there is no real hope for mankind, then so be it. In our fight against the forces of darkness, those sobriquets become tokens of distinction and we wear them with pride.

The book in front of you is not a learned treatise based on official studies, reports, and statistics. It is simply a conversational primer about mathematics. More specifically, it is a *sui generis*,[3] tongue-in-cheek discussion of how mathematics impinges on various aspects of human life—directly, through numeracy skills, or indirectly, through logical reasoning—combined with an attempt to highlight certain perceived inadequacies in our social universe that act as a

---

[1] Irish orator, philosopher, and politician, 1729–1797.
[2] American biochemist, author of science fiction and popular fiction, 1920–1992.
[3] Of its own kind; unique. (Latin)

brake on its progress and many of which, ultimately, may be blamed on the public's lack of proper education.

Why write on this topic? The reason can be traced back to two incidents, one old and one recent. The memory of the former has been with me for a long time, nursing my desire to speak out. The latter was the final trigger.

The old incident occurred when I was in second grade. One afternoon I went to a friend's house and asked him to come out and play. He said that he would, but only if I checked his math homework for errors. Since he went to a different school, I agreed and examined what he had scribbled in his notebook. At first I didn't understand anything; then, slowly, I began to grasp the enormity of what he had done: he had taken the numbers from the wordy problem ("If 6 farmers plow 10 hectares in 3 days, ...") and added, subtracted, multiplied, and divided them indiscriminately until he got the answer given in the back of the book!

At seven years of age, my options for dealing with this act of barbarism were very limited: I thought I could either try to make the offender see the error of his ways, or vent my outrage by beating the living daylights out of him. I chose diplomacy, but my peaceful approach was cut short with a shrug and an indolent "Aw, forget it, it's only math." *Only* math? One word, one single unfortunate word.... Fists erupted in all directions. Mathematical honor was satisfied and our friendship survived, but he never asked me to check his homework again. This was a defining moment for me because, for the first time in my life, I discovered how little some people understood of mathematics and how they misused and abused it.[4]

The recent incident was related to me by a colleague, whose neighbor had asked him how he could find the volume of a rectangular room. "It's easy," my colleague said. "Multiply the length and width of the floor, then multiply the result by the height of the ceiling." The neighbor was unhappy. "You are telling me how to find the *horizontal* volume, but I don't need that. I need the *vertical* one." Lucky for the neighbor, I suppose, that my colleague was not a seven-year-old.

The book is aimed at a wide audience. Practically anyone comfortable with polysyllabic English and logical thinking will derive some benefit from it. An enhanced benefit level is guaranteed for students who want or have to learn mathematics, teachers of the subject, professionals using it in their work, and anyone else interested in the fun side of this, the most noble of all sciences.

The contents are organized as a sequence of 25 SCAMs (Stories, Challenges, and Adventures in Mathematics) that encapsulate the essence of my conversations with an unflappable character by the name of J.J. Moon, whom I met at a number of conferences around the country. The first two SCAMs set the scene, introducing JJ and his *modus operandi*.[5] The rest are constructed on a com-

---

[4] I hasten to add that, as I grew up, both my temper and capacity for tolerance improved considerably.

[5] One's characteristic method of work. (Latin)

mon template, which begins with a couple of simple yet intriguing questions posed by JJ for students, followed by his pronouncements on an educational, cultural, ethical, or behavioral issue and my offering of two or three brief humorous stories. JJ's questions are then answered in a section titled *Notes After the Meeting*. Bringing up the rear, the section *A Word to the Wise* comments on a few well-known computational principles and lists DOs and DON'Ts that clarify their application.

The mathematical bits can be skipped without detriment to the free flow of the text, which is written in everyday language, uses the word 'like' only with its proper meaning, and has no heavy (or even light) philosophical undertones. Given that some readers may be young and unfamiliar with certain names and ideas mentioned in the book, I erred on the side of caution and provided lavish details in the solutions, succinct biographical information about the people whom I quote, and the translation of the foreign language words scattered (sparingly) throughout.

The problems and stories belong to mathematical folklore and are practically impossible to acknowledge. The best I can do in the circumstances is to thank my friends, colleagues, and students who have brought them to my attention by word of mouth or e-mail at one time or another, and the unknown (to me) sources from where they came. (I would be very pleased to hear from anyone who knows their true origin—not intermediate relays—so I could quote it in the future.) The final selection is mine and, inevitably, reflects my personal preferences.

A mathematics educator should teach not just scientific fact, but values as well; hence the chatty segments, which focus irreverently on certain aspects of the human condition and any link, solid or vague, direct or through analogy, they may have with logic and numbers. Of the recreational questions, some aim to show that not every solution is intuitively obvious, and some are designed to illustrate how one can blunder into all kinds of nonsense if rules are thrown out the window. The latter should give food for thought to all those who believe that mathematical felonies like division by zero have been decriminalized.

Do humorous stories have a place in mathematical discourse? Do they ever! According to Mark Twain,[6] humor stems from sorrow and there is no humor in heaven. He was right: humor is very much at home in the classroom, where we systematically torture our students with lectures and exams, pretending that we know what's best for them. If you listen to 50 minutes of definitions, theorems, and proofs thrown at you nonstop, your eyes will soon start glazing over; peppered with a few amusing lines, however, that abstract talk becomes almost bearable.

The formulas inserted at the end of the SCAMs are fundamental to correct mathematical manipulation and writing, and should be regarded as the bare essentials for students and sundry professionals who use mathematics as a tool.

---

[6] The pen name of Samuel Langhorne Clemens, American writer, journalist, and humorist, 1835–1910.

They are also a reminder for schoolteachers of what we require their graduates to know when they come to university. In my view, the individuals who can recite them without hesitation when woken up in the middle of the night have a very good chance to achieve a state of mathematical grace.

Much to my chagrin, in some parts of the book full rigor had to be sacrificed to help the nonspecialist understand the gist of the argument. I apologize in advance to those, including myself, who find this practice distasteful.

The book shines a satirical light on a lot of people and professions, but such banter is not meant to be disparaging or offensive. I must explain that J.J. Moon was a very trenchant and opinionated dialogue partner, the type who does not mince words and takes no prisoners. If he sounds condescending at times, it is because he is convinced that our way of life needs a few urgent upgrades, and that the quickest and most honest method to make people aware of this is to tell them the naked, unembellished truth. JJ's perception of the truth, which may be found unpalatable by some and infuriating by others, was pieced together from media reports and eyewitness accounts, often hyperbolized to drive the point home more forcefully. He advocates that you cannot bring about real change unless you burst complacency bubbles and throw harpoons at sacred cows. I tried hard, but in spite of my valiant efforts, he would not be deflected from his views. Consequently, I wish to make it absolutely clear that anything objectionable or controversial comes from JJ and has nothing to do with me. By contrast, anything considerate and pleasant is my merit alone.

You have been warned, now start reading the book—if you dare.

Christian Constanda
Tulsa, Oklahoma, July 2009

Acknowledgements

The writing of this book has been helped along by many people, whose suggestions, comments, and assistance with software resulted in significant improvements to the original draft.

Particularly deserving of my thanks are, in alphabetical order, Kimberly Adams, John Bailey, Christopher Banerjee, Karen Bouteller, Liliana Cazacu, James Childress, Bill Coberly, Peyton Cook, Bryce Culhane, Dan Constanda, Dale Doty, Smaranda Galis, Tom Grasso, Matthew Hale, Brook Iwata, Kyle Klavetter, Daniel Lazar, Elizabeth Loew, Kevin O'Neal, Mauricio Papa, Shirley Pomeranz, Eric Robison, Sujeet Shenoi, Melanie Smith, Kerry Sublette, Ian Tweddle, David Wallace, and Charles Windass.

No book sees the light of day without proper editorial guidance and management. These have been given to me unstintingly by Karen Borthwick at Springer–London, to whom I am grateful for keeping this project, firmly and efficiently, on a steady course from beginning to end. I also wish to thank Kathy McKenzie for her very thorough copy-editing of the manuscript, and Matthew Power and Lauren Stoney for their smooth handling of the production process.

My deepest gratitude, however, is reserved for my wife, Lia Constanda, whose logical mind, sound and thoughtful advice, remarkable patience, and unselfish support keep my fires burning bright.

Contents

Preface vii

Acknowledgements xi

List of Acronyms and Abbreviations xix

SCAM 1
JJ and the MICQ 1

Enter JJ. His persona, mission, and method. The MICQ: an accurate measure of selfhood. Succinct discussion of humans with a MICQ below the all-important mark of 50.

SCAM 2
The Mathematical Commandments 9

BRITEs and unBRITEs. The ten mathematical rules that underpin numeracy and their connection with the MICQ. A matter of logic. Baseball and cricket in an alien purgatory. How to put out a fire in the faculty lounge. Strange algebraic goings-on.

A word to the wise about logic and human behavior.

SCAM 3
The Public School System 19

Matching socks and contributing fairly to the cost of a party. The ills of public school education. Mediocrity, dissatisfaction, classroom discipline, and the teaching unions. A specimen test for graduating high school students. Salary is inversely proportional to knowledge. The wall of Jericho. Mass defection of mathematics professors to plumbing.

A word to the wise about equality and implication.

SCAM 4
The School Mathematical Education 33

A cowboy's testament. The consequences of falling in love with royalty. The Principles of Least Headache, Unlimited Confidence, Convenient Choice, and Wishful Thinking. New Whole Math. How to con the government out of $5 million. The motivational powers of the 'plus' sign. A snapshot of school mathematics through the decades.

A word to the wise about exact and approximate equalities.

SCAM 5
Language, Grammar, and Punctuation 43

Flies that demonstrate a mathematical point. About irrationality (in number land). What *its*, *like*, to rough up the English language. A way to kick the habit. Dangling participles are in, the Saxon genitive is out. Algebra, logic, and linguistic patterns. 'Logical punctuation', not 'illogical punctuation.' A typical teacher's day in the classroom? Spelling as a cure for ice cream obsessions. Professionals and the primes.

A word to the wise about the number system.

SCAM 6
Foreign Languages 55

Mathematical proof that global warming is way out of hand. Counting brothers and sisters. The mangling of foreign names and the tragedy of the Greek alphabet. Steak with *au jus*, among other things. One-digit clock faces. A polyglot lost in Australia. Of mice and languages.

A word to the wise about complex numbers.

SCAM 7
Foreign Countries and Foreigners 67

The growing habits of water plants. Chess and wheat don't mix. Geography: Achilles's heel of public education. Traveling oafs who bring the country into disrepute. To come or not to come to America. A mathematician's rejection of burial in the Holy Land. Do car brand names mean what they say? The life-saving qualities of geometric progressions.

A word to the wise about algebraic operations.

SCAM 8
Mathematics and the Public 79

Counting rounds in a competition. A dilemma over money in envelopes. Short-changing the second millennium. A crook who knows fractions. Win $4 million with polynomials (if the judge is prepared to learn what they are). Santa's visit on Halloween day. Help for a student with a drinking problem. How to lose a bet by underestimating your waitress.

A word to the wise about writing fractions.

SCAM 9
The Question of Calculus 91

Some farming business and proportionality. Can a tortoise outrun a hare? The essence of calculus and its traditional and reformed versions: dissension in the ranks. Fumbling across a cluttered room

Contents   xv

in total darkness. Reformed logic and the Principle of Truth by Close Proximity. Reformed calculus in the afterlife. Trying to outsmart the professor is not all it's cracked up to be.

A word to the wise about operations with fractions.

SCAM 10
Political Correctness 101

True assertions that cannot be proved to be true. Classifying individuals alphanumerically. Men, women, and dogs on an oil rig. Order of exit through a door. A porcine male chauvinist. Feminism in modern society. The purported evil nature of girlfriends. Fate of a deluded frog. Slowism and other bridge-crossing issues.

A word to the wise about algebraic identities.

SCAM 11
TV Advertising 111

Father and son: an age problem. Cracking numbers algebraically. Medicines, fast food, cell phones, and dog dinners. Confusion over the North and South Poles. The illusion of cheap high-speed internet rates. Disqualification through overqualification. How to sell insurance policies. Acquiring knowledge the easy or not so easy way.

A word to the wise about quadratic polynomials and equations.

SCAM 12
Highway Driving 121

Mathematical proof that a large part of the population is insane. Folding the state of Virginia over and over. Public reaction to incompetents behind the wheel. A driving test for dimwits? Weight versus centrifugal force: a matter of life or death (allowing for voluntary maiming). The hypocrisy about road surveillance cameras. A cab ride in Moscow. Old age doesn't necessarily mean decrepitude. The similarity between frying eggs and driving cars.

A word to the wise about square roots.

SCAM 13
Units of Measurement 133

Money has no value. Mathematical proof that the universe does not exist. Metrication: a vexatious issue. Positive integers superior to fractions at producing more and better plumbers. The only efficient change is sudden change. Driving on the left came first. Do pint glasses help publicans cheat their customers? The speed limit in France. Students' inconsistent demeanor before the finals.

A word to the wise about factoring and roots of equations.

SCAM 14
Percentages and Living on Debt 145

Mathematical proof that Earth is flat. Three sisters and a house number. The curse of percentages. Raises that are not what they seem to be. A lose–lose salary conundrum. Thoughtless borrowers do it to themselves. The fate of a credit card banker who takes the word of a dumb machine over that of an alien. A much wiser and quick-thinking banker. The Devil's most reprehensible act.

A word to the wise about percentages.

SCAM 15
Modern Art 157

How to sort out a restaurant check. A magic box that works on elementary algebra. Atonal music on the rack. Mathematical illustrations of modern painting and abstract poetry. Art, the golden ratio, the Fibonacci numbers, and fractals. Functions mauled by operators in a normed space. Transylvanian wedding mysteries. A Picasso aficionado's unusual date.

A word to the wise about exponents.

SCAM 16
Averages and Buying Cars 171

Mathematical proof that anyone can afford a Porsche. Who dates more: men or women? How to push a car salesman to within an inch of a null profit margin. Serpents' propagation secret. Computer scientists should know better than to challenge mathematicians' power of imagination. A bizarre kind of induction in the sky over the Atlantic.

A word to the wise about logarithms.

SCAM 17
The Public Media 183

Mathematical proof that realtors cannot be trusted. Cats crawling under long pieces of string. The high nuisance-index of some journalists. Accuracy in reporting and movie-making: a disposable inconvenience. The anticulture of sensationalism and deliberate suppression of credibility. Defeating galaxy-crossing aliens with a laptop. A professor's tennis ball. From hero to villain in the blink of a camera. Humor by numbers.

A word to the wise about geometry.

SCAM 18
The Criminal Legal System 195

How to tell judges and defense attorneys apart through pure logic. Mathematical proof that the law is an ass. Smart attorneys and not-so-smart jurors as slayers of truth and justice. The defense argument: a plus or a minus? Non-Euclidean geometry and the Constitution.

Contractual agreement between burglar and victim. Justice in JJ's society. The unhappiness of lawyers who ask witnesses the wrong questions. Math teachers have long memories.

A word to the wise about trigonometry.

SCAM 19
Civil Litigation 209

Mathematical proof that God is stronger than the Devil. The sham of class actions. Suits and more suits: from the bizarre to the ridiculous. Unreasonable and ignorant jurors and the odd sensible judge who keeps things in proportion. University administrator: the oldest profession on Earth? How an engineer can improve life in Hell.

A word to the wise about inequalities.

SCAM 20
Statistics and Probability 217

The Friendly Shell Game and the Inept Crook. Why does a student end up more times at the mall than at the university in the morning? The infallible weather forecast. What happens to a company that fires one of its managers without knowing elementary probabilities. A bookie confused over shared birthdays. Statistics are not always correct and meaningful. A quick lesson in eastern European geography and complementary probabilities. Unusual survey answers. Averages and shooting ducks.

A word to the wise about statistics and probability.

SCAM 21
Academic Politics 227

The wisdom of thinking outside the box. Mathematical proof that all students fail all their exams and graduate *summa cum laude*. How to form a doctoral committee. Democracy among profs. Why colleagues sabotage colleagues in reference letters. Walking on water defeated by mediocrity and cronyism in internal promotions. Allegiance to one's dean: the shark test. The importance of choosing the right dissertation advisor.

A word to the wise about limits.

SCAM 22
Antisocial Behavior 239

Mathematical little beasties and their deadly encounters with locomotives. A neat card trick backed by the logic of numbers. Loud music, abandoned pushcarts, untidy shoppers, yapping dogs, and some simple arithmetic. Cell phones and their ringing tunes. Getting ideas from a tiny mechanical rat. Alligators with square-shaped bodies.

A word to the wise about differential calculus.

xviii   Contents

SCAM 23
Mathematicians Versus Engineers 251

> Induction revisited. Proof that all engineering mathematics is incorrect. Verbal sparring between a mathematician and an engineer. A difference in methodologies. The strange way in which mathematicians give directions (especially to engineers). How to travel cheaply by train. The limitations of a genie's powers. Why gum-chewing engineers should not allow mathematicians to buy them drinks on a plane.
>
> A word to the wise about integral calculus.

SCAM 24
The Evolution of Knowledge 261

> Mathematical proof that physical sciences on Earth are a farce. Counting ad infinitum. The hierarchical quality of scientific theories. What a civilization without numbers does when it starts discovering them. The pole of learning: from Newton's mechanics to Einstein's theory of relativity and beyond. Mathematical modeling as an indispensable tool for gathering knowledge. Application to horse racing. A lesson in light bulbs.
>
> A word to the wise about infinity.

SCAM 25
The Virtues of Mathematics 275

> Proof of the divine nature of mathematics. How to divide chocolates between siblings (when you can't eat them all). Defending mathematics in front of a jury of peers. Calculators and computers as execution instruments, not solution designers. The invisible yet decisive influence of mathematics on everyday life. Abstract thinking: a source of aesthetic pleasure. Dissension caused by the result of a simple sum. Evidence on behalf of the queen of sciences. Who rules Heaven? Hiring with arithmetic.
>
> A word to the wise about names of symbols.

Epilogue 289

Index 291

List of Acronyms and Abbreviations

Global Acronyms and Abbreviations

The acronyms and abbreviations listed below, defined in the first three chapters, occur throughout the text without any additional explanation of their meaning. They are gathered here for quick reference.

MICQ	Mathematical Intelligence and Character Quotient
Votsit	Validator of Objectivity and Truth and Synthesizer of Imagery from Thought
Brite	Benevolent, Resourceful, Intelligent, and Talented Earthling
Joe	Just an Ordinary Earthling
Gawd	Great Architect and World Designer
Gan	Any member of the Ganymedean survey team
Stu	Any student

B.L.Z. Bub is self-explanatory.

Local Acronyms and Abbreviations

Other name abbreviations are defined and used only in specific stories. They are formed from the first three letters of a word occurring in a person's status/job description or are acronyms of words in the text which describe the person's character. Examples:

Man	*Man*ager
Att	*Att*orney
Cas	*C*reative *a*rts *s*tudent
Sec	*S*illy *e*gregious *c*laimant

SCAM 1

JJ and the MICQ

> *There are more things in heaven and earth, Horatio, than are dreamt of in your philosophy.*
>
> William Shakespeare[1]
>
> *Have you had a close encounter recently? A close encounter with something very unusual?*
>
> Steven Spielberg[2]

"My name is J.J. Moon, and I'm an alien," he said calmly.

We were in the bar of a hotel in Auburn, Alabama, where I was attending a mathematics conference sponsored by Auburn University. The afternoon session had just ended, and I wanted to relax with a cup of coffee.

I had noticed him the moment I stepped into the room: he was sitting alone at a table in a corner, gazing at me with dark, inscrutable eyes. Almost unaware of what I was doing, I had walked over and joined him.

"You're a foreigner? Why would this be unusual?"

"No, I am an *alien*. A being from a different world."

Oh, dear, I thought. That was not the kind of company I needed before dinner.

"I'm a Ganymedean,"[3] he specified.

My long stare, which seemed to have no effect on him, convinced me that there was definitely something strange about the man, both disconcerting and intriguing. A sudden mixture of curiosity and fascination kept me in my seat.

"Really? And why are you here?" I asked.

"I'm attending this conference, like you." He pointed to a name tag pinned to his shirt and inscribed 'J.J. Moon, USENC'.

"Some university in North Carolina?"

"Very few people bother with the initials. But when they do, this is what they normally guess, and I don't contradict them. In fact, the letters don't designate an affiliation. They stand for Unwavering Supporter of Education, Numeracy, and Culture. A little game of mine. Anyway, how did you like the last talk today?"

---

[1] English playwright and poet, 1564–1616. From *Hamlet*.
[2] American director and screenwriter, b1947.
[3] Ganymede is a satellite of the planet Jupiter. In Greek mythology, Ganymede was a Trojan boy who became cupbearer to the gods.

"It was surprisingly good. I expect to see a whole new class of methods come out of that nice piece of work."

"Right," JJ said neutrally.

"So what are you, a Ganymedean, doing on Earth?" I asked again, trying to get back to what we had been discussing before: aliens and stuff.

"Long story," he said. "Our home was the largest Jupiter satellite, until we moved it into a space fold that we created ourselves near Ganymede. From that hiding place we have traveled far and wide, visiting and studying many other worlds, including yours. Unfortunately, we cannot perform all the tasks here by walking about in our original bodies. For one thing, your atmosphere is harmful to us, chemically and biologically; and for another, we must not reveal our presence to you. However, through sheer luck, our science has found a way to morph Ganymedeans with a certain type of genetic makeup into plausible humans. I'm one of the volunteers. As you can see, the change is not perfect, but it's good enough to make us practically indistinguishable from the natives."

I looked at this Moon character more closely: short, thin, with large black eyes and skin of a subtly iridescent, dark olive hue. Yes, I thought, a regular little green man from outer space. Except that he seemed fully human in every respect. I decided to humor him a bit longer before making my excuses and leaving.

"I take it that you don't remain in this morphed guise forever. When you go back to your ship you do revert to your original self, don't you?"

"The morphing process is complicated and delicate. Our bodies would sustain extensive damage if we went back and forth too frequently. On average, we keep our borrowed form for about five years at a time. Evidently, this forces us to blend into the population and live as close a life to a human's as we can. I have a house and a job like everyone else."

"Why not stay on the ship?"

"Have you ever tried to live aboard a flying saucer? I can assure you, it's very uncomfortable."

Well, then: here was the simple explanation of a mystery that had been baffling us for years.

"If what you say is true, then why tell me? Aren't you afraid that I might blow the whistle on you and your operation?"

He seemed to smile, infinite patience welling in his dark eyes. "It's part of my mission. I've just finished a long-term assignment in the U.K.—"

"Interesting coincidence," I cut in. "I've also worked there for a few years. Beautiful country, great history, terrible weather."

"—and I'm now looking for suitable U.S. individuals who will agree to listen to our message and pass it on to the population at large," JJ finished, ignoring my interruption.

"And I'm suitable?"

"That depends."

"On what?"

"On whether you are receptive to our efforts and willing to help us."

"Have you tried others before?"

"Many. Some made the grade, some not. Most didn't."

"Still, why me?"

"I thought that, since you like science fiction and are an intelligent person, you wouldn't be too upset to hear who I am, or too frightened to talk to me. As for blowing the whistle, well... you tell anybody that you met an alien and they'll call the men in white coats to lock you up."

I gasped. "How do you know that I like science fiction?"

"I'm a mind reader."

"No kidding," I said derisively.

"Don't worry, you are quite safe: I cannot 'plant' thoughts in your mind. I should also tell you that trying to arrange for electronic proof of my existence would be futile. I will not be recorded with what you've got on this planet. The tape or disk will remain blank, which will make you look even more foolish."

Yes, he could read minds. Mine, anyway.

"Fine, I believe you," I said, not really knowing what else to say. Then it struck me that the man had a sense of humor. I pointed to his name tag. "J.J. Moon: *Jovian* Moon?"

That *soupçon* of a smile again. "Correct."

"What's the other J for?"

"You see," he explained, slightly embarrassed, "we have developed this quaint little custom. When the first morphed Ganymedean was let loose on Earth, he didn't know what name to assume, so he decided to make it up from the first words addressed to him by a human. Thus, we have researchers called Y.A. Joking, Y. Sure, A.Y. Forreal, Y. Dontsay, P.T. Otherone, and so on. We also have someone called N. Shmit: the computer got it wrong and misspelled the name."

"And you are J.J. Moon because...?"

"Hearing where I'd come from, my contact looked at me and exclaimed, 'Jeeesus! The Jovian Moon!' So I became J.J. Moon."

I nodded. "Makes sense. But what do they call you on Ganymede?"

"!@#'!?&, if you must know."

"I don't think I can say that."

"Probably not. Unlike your language, ours has neither vowels nor consonants; just sounds."

"How shall I address you, then?"

"JJ will do."

"Want some coffee, JJ?" I asked him. His head motioned a clear refusal. I wondered briefly what Jovians morphed into humans liked to drink, if they became anything like us.

"I'm curious," I said, changing the subject. Strangely, I found it easy talking to him, whoever he was. "Tell me: what do you think about us, humans, compared to your species?"

"You aren't doing too badly, by and large, but have a very long way to go before you'll get where we are now."

I didn't need JJ to tell me that. "How will we know when we get there? When we've 'come of age'?"

"You'll be there when your planetary MICQ reaches 50."

"Our planetary *what*? Did you say 'mick'?"

"No, MICQ: M-I-C-Q. Short for Mathematical Intelligence and Character Quotient."

"Ah, well, we know all about IQs and how to measure them."

"There's a huge difference between the MICQ and your IQ," JJ said. "What you gather with your simplistic tests is a bunch of *finitely many* data, insufficient to evaluate a person's full worth. The MICQ is not just a reflection of an individual's dexterity with numbers, spatial vision, or ability to solve little puzzles. The MICQ measures a *continuum* of variables, collating their past and present values into a comprehensive picture of one's capacity for logical thinking, common sense, imagination, culture, and overall personality."

"We also have personality tests."

"Yes, you have both IQ tests and personality tests. But they are very primitive tools, and you don't know how to combine them into a coherent, multifaceted, and complete profile of the subject. You don't have the mathematical model for that, or the technology to implement the computational algorithm."

This caught my interest. "And you do?"

JJ looked as if he wasn't certain whether to tell me or not. "We do," he said after a few seconds. "Unfortunately, I cannot explain it to you. Even a professional like yourself wouldn't understand the details. Besides, the rules of engagement forbid us to disclose our advanced methods to alien civilizations. You must discover the process, step by painful step, on your own. What I can tell you, though, is that it's all based on very sophisticated mathematics—the only science capable of explaining everything in the universe, living creatures included. I'm giving away no secrets if I say that you, humans, have so far completed only a few crumbs of preliminary work; for example, you've modeled the functioning of neurons. Next, you need to start thinking how to construct a mathematical model for the entire brain. Once you master that, you can then tackle the most complex problem of all: the modeling of human personality. But to get there, you'll need computational power many orders of magnitude higher than what you've got now. The MICQ is calculated by feeding an individual's data into that ultimate model."

"It's still a mathematical index," I said.

"Yes," JJ agreed. "But its name should not be taken to imply that all it evaluates is just the strength of the mathematical component of a being's intelligence. The name means that the index is *constructed mathematically*, with the authority of full and impeccable rigor."

"And how do you measure all those infinitely many variables? Doesn't that take an infinitely long time?"

"We use a special gadget that scans the brain of the subject, reads everything imprinted on it, and dumps the reading into the analyzer. For your race, we have compressed the spectrum of measured essential features to the interval from 0 to 1 and calibrated the machine to output a result in the range from 0 to 100. The final figure, the MICQ, is the absolute minimum on the subject's profile curve. Here," JJ said. He produced a blank piece of paper and sketched a graph.

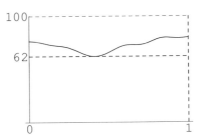

"A subject with this distribution of values has a MICQ of 62. As you can see, the profile curve does not change violently. We are dealing with a pretty well-balanced individual, who does credit to mankind."

Then he sketched another graph.

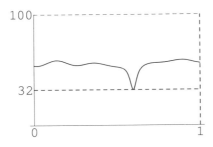

"This one would also be a fine specimen except for one major defect, which lowers his MICQ to 32. You might find such a profile, for example, in a politician. Whereas this one," he added, making a third sketch, "is what you expect to get from a vicious criminal."

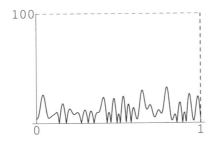

"If I understand you correctly," I said, "a person with an excellent mathematical mind and oodles of general knowledge may have a midrange, or even low, MICQ because his personality is badly flawed in some respect."

"That's right. One single objectionable character trait and down his MICQ goes. The converse is also true: someone with moderate mathematical knowledge but a great personality may have a high MICQ. The device does not measure only how numerate the subjects are; it also reads their capacity for understanding mathematics under proper instruction. In other words, it detects the presence of the necessary cerebral circuitry, even if it hasn't been activated. As you know, stupidity is absolute and incurable; ignorance, on the other hand, is relative and can be treated. The model is very intricate."

I looked him straight in the eyes. "Do you have the gadget on you? What do you call it in English?"

"Since it's a multipurpose instrument, and since your kind seems to be rather fond of acronyms, let's call it a VOTSIT—short for Validator of Objectivity and Truth and Synthesizer of Imagery from Thought. And yes, I do have one on me."

"Can I see it?" I said, a little too eagerly.

"Absolutely not."

"Then why should I believe you? The things you've been telling me are so extraordinary that without hard proof—"

"Oh, you want proof! Tell me again: how did you like the last talk this afternoon?"

"A total waste of time. The speaker showed nothing new, and what he'd done would never be used in any practical way. A dead-end piece of redundant research, if you ask me." Even as I let those words out, I could not believe my ears.

"Earlier, you said that—"

"I know what I said," I snapped nervously.

JJ tilted his head. "The device makes the subject tell the truth, among other things. It records and plays back interviews. It also converts images formed in the subject's mind into audio-visual projections. We use it all the time in our surveys."

I took a deep breath and calmed down. "Since we don't have a VOTSIT to make an accurate determination of one's MICQ, can we at least get an idea of its approximate value? Is there a rule of thumb for that?"

"We've classified humans into five-point bands according to their MICQs."

"But you cannot tell me what they are."

"I don't see why not. As I said earlier, the threshold for mankind to be considered a 'mature' race is 50, so here is how the MICQ makes placements in the grand scheme of things up to that level, in terms of human values and other factors. For each band I'll mention an example of typical personality flaw and a couple of groups of people whom we found to belong there. I emphasize that the list is not exhaustive by any means."

Bringing out another piece of paper, JJ started to scribble rapidly a summary in the form of a table.

MICQ	Sample Personality Flaw	Sample Membership
0	inferior human material	hard criminals unrecoverable drug addicts
1–4	acute selfishness	reckless highway drivers antisocial individuals
5–9	mindless greed	irresponsible heavy debtors house-trained dogs
10–14	bigotry	perpetrators of discrimination incompetent jurors
15–19	compulsion to fool the gullible	artful car salesmen amoral media zealots
20–24	dearth of inspiration	untalented TV advertisers peddlers of political correctness
25–29	tendency to distort the truth	defense lawyers language polluters
30–34	self-serving demagoguery	most politicians specious educationists
35–39	delusions of grandeur	pseudo-intellectuals phony 'modern' artists
40–44	Philistinism	spurners of general culture opponents of metrication
45–49	inability to grasp abstract concepts	undistinguished engineers roguish statisticians
50	lack of sense of humor	rigid military commanders inflexible tax auditors

"Would you care to elaborate?" I asked, glancing at his table.

"I can't explain everything in one go. All I'll say for now is that 0 is automatically ascribed by the MICQ compiler to murderers, rapists, child molesters, armed robbers, cruel and violent individuals, drug dealers, and substance abusers beyond redemption. These specimens have put themselves outside the normal parameters of the human race, and I will never mention them again."

"Why did you include dogs in the 5–9 band? I thought the classification applied exclusively to humans."

"A colleague of mine accidentally used the VOTSIT on a neighbor's pet poodle and was shocked to find that it gave a small positive reading. We then experimented with other canines and got similar results. Our theory is that dogs, when properly trained, do not upset things in the house, obey simple commands, and are loyal to their owners—a clear sign of rudimentary intelligence."

"What about the upper half of the MICQ scale? Who would, for example, have a MICQ of 60?"

"A professor who is honorable in every way but treats his students like second-class citizens."

"I'm glad that you qualified the 45–49 entrants. The vast majority of engineers and statisticians respect mathematics and apply it well."

"You're right, although 'vast majority' may be a slight exaggeration. There are also qualifications on other subgroups, as I'm sure you've noticed."

"Why not on defense lawyers?"

"We haven't come across any exceptions yet."

I had a whole bunch of questions to ask JJ regarding his MICQ-based hierarchy, but I didn't get the chance. Suddenly, he looked at his watch and said, "I've really enjoyed our little chat. If all goes well, we'll talk some more next time." Then, unceremoniously, he sprang to his feet and promptly walked out of the bar.

I tried in vain to find him during the last day of the convention. He had simply vanished. Or maybe he hadn't been an attendee in the first place? Who was he? A prankster? A nut? As Rick Blaine[4] might have said, of all the conferences in all the towns on all the worlds, he had to come to mine. Could he really be who he'd said he was? Anyway, I didn't think I'd ever see him again.

But I was wrong.

---

[4] One of the main characters in the film *Casablanca*.

SCAM 2

The Mathematical Commandments

> *A few strong instincts and a few plain rules suffice us.*
> Ralph Waldo Emerson[1]
>
> *Die Wissenschaft kennt nur ein Gebot: den wissenschaftlichen Beitrag.*[2]
> Bertolt Brecht[3]

I met JJ again at a mathematics conference sponsored by Brown University in Providence, Rhode Island. Just like the first time, he was sitting alone at a table in a corner of the hotel bar, gazing at me across the room with dark, inscrutable eyes.

"Civilizations rise and fall on their treatment of mathematics," he said abruptly when I joined him.

"What's brought this on?"

"I've just completed an in-depth analysis of the level of basic mathematical knowledge on Earth and concluded that it's quite unsatisfactory."

"Tell me something I don't know."

"I've also condensed the remedy into 10 simple principles—kind of commandments, you might say—that should be brought to the attention of all students. If these commandments are obeyed, your planetary MICQ will improve dramatically in short order and start approaching the critical value of 50."

"Who do you think you are: the Almighty?"

JJ did not seem to take offense. "I guess you're referring to the Great Architect and World Designer," he said. "In that case, no, I'm not GAWD. These are truths that should be obvious to any sentient creature who operates with number-based mathematics."

"You've told me about a person's MICQ, but not how the *planetary* MICQ is computed."

"As you'd expect, the planetary MICQ is a type of average. However, its calculation is highly nontrivial because, to be realistic, the final score must reflect the existing balance between the good and the bad, between the terrible deficiencies and the remarkable achievements of your race, between the baggage you are carrying from the past and the promise you are showing for the future."

---

[1] American essayist, poet, and philosopher, 1803–1882.//
[2] Science knows only one commandment: contribute to science. (German)//
[3] German poet and playwright, 1898–1956.

"So where do we stand at the moment?"

"The current global value is 27."

I felt deflated: I had hoped for a much higher figure. "You mean, we are still some sort of bipedal amoebas—"

"In our classification, a human is either a BRITE—a Benevolent, Resourceful, Intelligent, and Talented Earthling—or an unBRITE. The conventional line between the two is drawn at MICQ-50, the same as for the planetary mark. Ganymedeans like me talk to people like you to help accelerate desirable improvements in your species, to shrink the set of unBRITEs and enlarge the set of BRITEs. We hope that you will try to persuade as many JOEs as possible to mend their ways and change sides."

"JOEs?"

"JOE: Just an Ordinary Earthling," JJ clarified.

How many acronyms was he going to use? "And you think that ten mathematical rules are sufficient? I could've come up with at least a hundred."

"Ten will do. You need only the most basic ones. Everything else follows from there."

"Okay, show me the list."

I thought that JJ had prepared a written text. He hadn't. He simply launched himself into a careful enunciation, ticking the numbers off on his fingers.

1. Thou shalt not divide by zero.
2. Thou shalt not cancel common additive terms between the numerator and denominator of a fraction.
3. Thou shalt not make division distributive over addition.
4. Thou shalt not think of infinity as a number.
5. Thou shalt not hold that every function behaves linearly.
6. Thou shalt always be guided by logic in thy thinking and actions.
7. Thou shalt denounce the evil of reformed calculus.
8. Thou shalt use a calculator only when thy brain is not fast enough.
9. Thou shalt proclaim the precedence of mathematics over everything.
10. Thou shalt look with healthy suspicion upon engineering mathematics and the conclusions of statisticians.

"The first five," JJ said, "are prohibitions. They cover what may be perceived by the unwary as computational misdemeanors but which are, in fact, serious mathematical crimes. The rest are affirmative directives. No explanation is necessary for the sixth one. The seventh commandment draws attention to a malady that must be eradicated before it becomes endemic and does untold damage to mankind. The eighth one reinforces the truth, ignored by many, that electronic devices do nothing more than provide speed of execution for operations we understand and can program correctly. The ninth calls for recognition of the undisputed supremacy of mathematics in the world of science. The tenth statement is self-explanatory."

An interesting list, I thought. "What happens if one respects all these commandments?"

"Then his MICQ is 50 or higher. Unless he has other nasty skeletons in the personality cupboard to sink his number."

"Have you thought of a mathematical equivalent to your earlier MICQ classification, in terms of these commandments?"

"Definitely," JJ confirmed. "Here is a sample of tenets that some people embrace and cling to, even after going through proper mathematical education, together with the mathematical commandment they are most likely to contravene and the MICQ band where the misdeed places them." And he jotted down the following table.

MICQ	Offending Belief	Violation
0	Math is unimportant and should be outlawed	N/A
1–4	What involves nothing yields nothing	1
5–9	If it's exactly the same, zap it	2
10–14	Mathematical rules deny freedom of expression	3
15–19	Size, whether large or small, is always a number	4
20–24	All mathematical operations have equal rights	5
25–29	Gut feeling and whim beat pure reasoning any day	6
30–34	When something works well, change it	7
35–39	What a human can do a machine can do better	8
40–44	Math plays only a minor part in society's progress	9
45–49	Rigor is an expensive and unnecessary luxury	10
50	Knowing numbers means knowing math	none

"What MICQ would you ascribe to a person who transgresses more than one commandment?" I asked, pointing at the paper.

"The value corresponding to the lowest number."

"You made division by zero the most damning. Why?"

"Because it cannot be performed," JJ said. "If you look at the illegal operations indicted by commandments 2–5, you'll see that there are *particular* cases where the proscribed aberrant handling yields the correct result. Division by zero, on the other hand, is simply inconceivable."

I swirled the coffee idly in my cup. "Any suggestions how we can teach mankind more efficiently to abide by these ordinances?"

"Start with the sixth, which has implications far beyond mere mathematical thinking. Many individuals, from heads of state to ordinary folk, have been observed to behave illogically—they should be your primary target."

"But logic is just a bunch of principles governing correct reasoning and inference. As such, it's not unique. How, then, can we judge if the actions of others are correct or not? By which logic do we arbitrate?"

"By what I would call 'standard logic'. The logic affiliated with common sense and employed by the majority of normal, intelligent, educated, and discerning humans in any given situation, after every possible alternative has been judiciously considered. Suppose, for example, that a person has reached a canal and needs to get to the other side. She glances to her right and left and sees a bridge 200 m away. Standard logic dictates that she walk to the bridge and step over to the opposite bank. She shouldn't jump into the water and swim across—that would be illogical."

"Unless," I countered wryly, "she is being charged by a hungry lion, in which case taking the plunge is the recommended solution."

"A logical response certainly depends on the prevailing circumstances. Anyway, we, Ganymedeans, believe that there is no better grounding in the fundamental rules of logic than an elementary mathematical education. Illogic and innumeracy are strongly correlated."

"Are you saying that mathematicians are never illogical?"

"Not at all. Even a brilliant mathematical mind does not guarantee a fully logical demeanor. Remember Évariste Galois?"[4]

"The 21-year-old genius who got himself killed in a duel over his infatuation with a woman."

"One of the few exceptions that confirm the rule. This foolish young man's fate notwithstanding, mathematical instruction, even at a very basic level, remains the best method for training your race in the ways of sound and reliable argumentation, and, therefore—it is hoped—conduct."

"Do all Jovians obey the ten mathematical commandments?"

JJ recoiled. "Of course. Our planetary MICQ is well above 50."

"But this doesn't mean that *every* Ganymedean has a MICQ above 50. You said so yourself. Don't you, from time to time, come across someone who tries to divide by zero?"

The olive tone of JJ's face seemed to pale. "Umm...yes. But very, very seldom."

"So what do you do with the offenders?" I insisted.

JJ averted his eyes and, in a hoarse whisper thick with terror, said, "Callisto."

"That's another one of Jupiter's satellites. What of it?"

"By law, these rare deviants, and anyone else whose MICQ is lower than 50, are given a choice. They may accept to undergo brain-reformatting, a procedure that eliminates the existent defect by 'smoothing out' their entire

---

[4] French mathematician, 1811–1832.

personality and transforming them into bland, unexciting beings; or they are deported to our penal colony on Callisto."

"Much more civilized."

"On the contrary. Those sent to Callisto are forced to watch, day and night, baseball and cricket matches recorded on Earth."

A shiver coursed down my spine. "This is brutal!"

"Totally unbearable."

"How long do they have to remain there?"

"Until they learn to describe their feelings accurately," JJ said. "Which means forever, since the Ganymedean language doesn't have words for 'excruciating' and 'boredom'. A majority of those who go to Callisto never resist for more than a few months. Sooner or later, they beg to have their sentence commuted to brain alteration."

"Going back to this intriguing MICQ concept: you named the lack of a sense of humor as one of the most costly demerits for those in the 50-band. Why is sense of humor so important?"

"Humor makes life appear more exciting. It puts people in a better frame of mind, conducive to good quality work and healthy social interaction. It sharpens the wit and stimulates lateral thinking."

"Is it required of mathematicians, too?" I asked.

"Particularly of them. Mathematics is beautiful, but very abstract. You need to spice things up a bit when you talk about it in the company of others. In the classroom, you must teach it humanely."

"Do Ganymedeans tell jokes?"

"We do. But they are a little different from the staple Earth jokes. They usually involve science or scientists and may not appeal all that much to JOE."

"Give me an example, please."

JJ shrugged. "Okay, here is one."

Three friends—a biologist, a physicist, and a mathematician—sit in a pub, watching the world go by on the street outside. At some moment, they see a man and a woman enter the building across the road. A few minutes later, the two come out of the building accompanied by a third person.

"I can explain that," the biologist says. "They multiplied while they were inside."

The physicist disagrees. "Not enough time. You've had too many beers. This must be a measurement error." He turns to the mathematician: "What do you think?"

"I think," the mathematician says, "that if now exactly one person enters the building, it will become empty again."

"And here's another," JJ added before I had time to comment.

A physicist and a mathematician sit in the faculty club, sipping coffee and reading newspapers, when, suddenly, the coffee machine bursts into flames. The physicist gets up, grabs an empty bucket from under the sink, unplugs the machine, then fills the bucket with water and douses the fire.

As it happens, a week later the same physicist and the same mathematician are again in each other's company in the club, and again the coffee machine malfunctions and flames up. Noticing this, the mathematician calmly gets out of his chair, grabs the bucket, walks over to the physicist, puts it in his hands and tells him, "I have reduced the problem to a preceding case, which is already known to have a solution."

"Yes," I said, "your stories do indeed encapsulate something specific to mathematical logic and method."

"I'm collecting jokes that involve mathematics and mathematicians," JJ said. "So, if you know any, please pass them on to me."

"Why don't you look around yourself? There must be thousands of them out there."

"I don't have time to sift through all that material. I'd rather you select for me what you, a human, think are the best."

"I'll see what I can do," I answered noncommittally. "But I need something in return."

"What do you have in mind?"

"I don't know. I'll leave the choice to you. And by the way, I meant to ask: what's your own MICQ score?"

"That's none of your business."

"Come on, JJ! I won't be shocked, or intimidated, however high it is."

JJ relented. "All right. If you really must know, here on Earth my MICQ is 87."

"Why not higher? What's your worst sin, the one that drags you down to 87? And why did you say 'here on Earth'?"

"In my present incarnation, I've got an insatiable craving for watermelons."

I managed to keep a straight face. "That's awful! You lose 13 points because you like those cool, sweet, juicy suckers?"

"It doesn't matter how inoffensive it is. It's still an addiction, and every addiction gets penalized. You want to know what your worst sin is?"

Suddenly, my curiosity vanished. But I didn't have much choice, so I nodded. JJ's answer surprised me.

"You expect your students to actually *study* during the semester. You are too intransigent."

"So what's my MICQ, then?"

"Look," JJ said, pointing to his watch. "It's getting late. I've really enjoyed our little chat. If all goes well, we'll talk some more next time." Then, unceremoniously, he sprang to his feet and promptly walked out of the bar.

Notes After the Meeting

Using basic algebra, we easily find particular cases where, as JJ pointed out, the operations banned by rules 2–5 on his list happen to produce the correct result. Take the Second Mathematical Commandment, which decrees that, for

example,
$$\frac{a+b}{a} \neq \frac{\not{a}+b}{\not{a}},$$
in other words, that, in general,
$$\frac{a+b}{a} \neq b.$$
But if $a = 1.5$ and $b = 3$, the two sides above are equal:
$$\frac{a+b}{a} = \frac{1.5+3}{1.5} = \frac{4.5}{1.5} = 3 = b.$$

In fact, we can compute *all* pairs (a, b) of real numbers for which this occurs. Treating the equality
$$\frac{a+b}{a} = b \qquad (2.1)$$
as an equation for $a \neq 0$, we solve it to find that $a + b = ab$, or $ab - a = b$, from which
$$a(b-1) = b,$$
with solution
$$a = \frac{b}{b-1}, \quad b \neq 0, 1.$$

(The operations performed above mandate the restriction $b \neq 0, 1$, which also guarantees that $a \neq 0$, to avoid a breach of the First Mathematical Commandment.) However, the existence of such numbers a and b does not make (2.1) universally valid. What we have shown here is that (2.1) holds on the set
$$\left\{ (a,b) : a, b \text{ real}, \ b \neq 0, 1, \ a = \frac{b}{b-1} \right\}$$
and not for all a and b with $a \neq 0$.

The same is true about the more general equality
$$\frac{a+b}{a+c} = \frac{b}{c},$$
valid only for $a = 0$, $c \neq 0$ or $b = c$, $c \neq 0, -a$.

Similarly, the operation
$$\frac{a}{b+c} = \frac{a}{b} + \frac{a}{c}, \qquad (2.2)$$
forbidden by the Third Mathematical Commandment, holds for a particular set of triples (a, b, c). To find this set, we assume that b, c, $b + c \neq 0$ (to avoid division by 0) and multiply every term in (2.2) by $bc(b+c)$. As a result, we obtain
$$abc = ac(b+c) + ab(b+c) = abc + ac^2 + ab^2 + abc.$$

Canceling one of the terms abc on the right-hand side with that on the left-hand side and factoring out a among the remaining three terms, we arrive at

$$a(b^2 + bc + c^2) = 0,$$

from which $a = 0$ or $b^2 + bc + c^2 = 0$. The first alternative means that (2.2) is valid on the set

$$\{(0, b, c) : b, c \text{ real}, \ b, c, b + c \neq 0\}.$$

The second can be rewritten in the equivalent form

$$\left(b + \tfrac{1}{2} c\right)^2 + \tfrac{3}{4} c^2 = 0,$$

which does not hold for any pair (b, c) of nonzero real numbers (since the sum of two positive numbers cannot be 0). However, if we extend our considerations to complex numbers (see p. 65), we find that

$$b = \tfrac{1}{2}\left(-1 \pm i\sqrt{3}\,\right)c, \text{ where } i^2 = -1.$$

Therefore, we conclude that (2.2) is also valid on the set

$$\{(a, b, c) : a, b, c \text{ complex}, \ c \neq 0, \ b = \tfrac{1}{2}\left(-1 \pm i\sqrt{3}\,\right)c\}.$$

Regarding the Fourth Mathematical Commandment, all I would say at this point is that a symbolic (*not* numerical) 'operation' such as $\infty + \infty = \infty$ can sometimes be tolerated.

The Fifth Mathematical Commandment does not deny the existence of linearly behaved functions; it merely states that not all functions behave linearly. Relaxing the definition,[5] we say that a function f behaves linearly if

$$f(x + y) = f(x) + f(y), \quad f(cx) = cf(x) \tag{2.3}$$

for any x, y, and c for which all the terms in (2.3) can be computed. It is easy to see that every linearly behaved function of one variable is of the form

$$f(x) = ax, \quad a = \text{constant}.$$

However, there are sets of pairs (x, y) and (x, c) for which some functions with nonlinear behavior satisfy at least one of the equalities (2.3). Here are a few trivial cases. (For simplicity, we confine ourselves to real functions.)

(i) The 'square root' function $f(x) = \sqrt{x}$ satisfies the first equality (2.3) if

$$\sqrt{x + y} = \sqrt{x} + \sqrt{y}, \quad x, y \geq 0,$$

---

[5] The full and correct definition of a linear mapping can be found in any good book on linear algebra.

which, on squaring both sides and canceling the like terms x and y, yields the set
$$\{(x, y) : x, y \text{ real}, x = 0, y \geq 0 \text{ or } x \geq 0, y = 0\}.$$

(ii) The logarithmic function $f(x) = \ln x$ (see p. 181) satisfies the first equality (2.3) if
$$\ln(x + y) = \ln x + \ln y = \ln(xy), \quad x, y > 0.$$
Exponentiating on both sides, we find that this is equivalent to
$$x + y = xy,$$
which leads to
$$x = \frac{y}{y - 1}, \quad y \neq 1.$$
Hence, the desired set is
$$\left\{(x, y) : x, y \text{ real}, x, y > 0, y \neq 1, x = \frac{y}{y - 1}\right\}.$$

(iii) The 'square' function $f(x) = x^2$ satisfies the second equality (2.3) if
$$c^2 x^2 = cx^2, \quad c \geq 0.$$
This is equivalent to
$$c(c - 1)x^2 = 0,$$
or
$$c = 0 \quad \text{or} \quad c = 1 \quad \text{or} \quad x = 0.$$
Thus, we arrive at the set
$$\{(x, c) \text{ of the form } (x, 0) \text{ or } (x, 1), x \text{ real, or } (0, c), c \geq 0\}.$$

One important thing should be very clear: when we say that a mathematical operation has a certain property, we automatically imply that the property in question holds for all the members of the set where the operation is defined, and not just for some. Thus, since
$$a(b + c) = ab + ac$$
for *all* numbers a, b, c, whereas the equality
$$\frac{a}{b + c} = \frac{a}{b} + \frac{a}{c}$$

is true only for *some* numbers a, b, c, we decree that multiplication is distributive over addition, but that division is not.

Finally, let us justify the First Mathematical Commandment. If we think of division as the inverse operation to multiplication, then

$$\frac{a}{b} = c$$

should be equivalent to $bc = a$. The latter, however, does not hold when $a \neq 0$ and $b = 0$, so division by zero is not defined.

Alternatively, we may regard division as repeated subtraction. For example, we say that

$$17 \div 5 = 3 \text{ remainder } 2$$

because

$$17 - 5 = 12, \quad 12 - 5 = 7, \quad 7 - 5 = 2,$$

with 5 subtracted consecutively three times, that is, until the result is a nonnegative number less than 5. But this procedure doesn't work when 5 is replaced by 0, since 0 may be subtracted from 17 any number of times and the result will always remain 17, which is strictly greater than 0. So, again, we must conclude that division by zero cannot be defined in ordinary arithmetic.

A Word to the Wise

Logic and Human Behavior

In traditional academic interpretation, logic is the science that studies the construction of statements and arguments. Its various branches include informal, formal, symbolic, and mathematical logic, but this is neither a completely clear-cut, nor an exhaustive, classification.

When we attempt to apply logic to human behavior, the picture is uncertain and equivocal. People's attitudes and actions are normally judged by reference to an expected and accepted pattern, which is not always readily defined. Hence the need for the 'reasonable majority' principle that explains what JJ called standard logic. Individuals who operate on the basis of a different system of reasoning and violate this principle would be considered illogical by the standard-logic users. Here are two impromptu examples of such alternatives.

(i) The 'ostrich logic', predicated on the belief that what you don't see, can't harm you. Someone crossing a street without looking at the traffic would quickly be disabused of the notion that he is acting logically.

(ii) The 'vandal logic', underwritten by the conviction that if it's not yours, you may trash it. It is this type of logic that produces roadside litter, graffiti, and unusable public restrooms.

As social beings directly affected by the state of our own health and environment, it seems that we, humans, would do well to stick with the principle of reasonable majority and its standard-logic companion.

SCAM 3

The Public School System

> *The aim of public education is not to spread enlightenment at all, it is simply to reduce as many individuals as possible to the same safe level, to breed and train a standardized citizenry, to put down dissent and originality. That is its aim in the United States, whatever the pretensions of politicians, pedagogues and other such mountebanks, and that is its aim everywhere else.*
> Henry Louis Mencken[1]

> *True education makes for inequality; the inequality of individuality, the inequality of success, the glorious inequality of talent, of genius; for inequality, not mediocrity, individual superiority, not standardization, is the measure of the progress of the world.*
> Felix Schelling[2]

I met JJ again at a mathematics conference sponsored by Cornell University in Ithaca, New York. Just like the first time, he was sitting alone at a table in a corner of the hotel bar, gazing at me across the room with dark, inscrutable eyes.

As soon as I sat down, he pushed a piece of paper in my direction. "Here," he said laconically.

"What's this?"

"For your students. My part of the bargain. In exchange for the slice of humor you'll serve me today."

I glanced at the paper and smiled. The text read something like this.

Question 1 (requires common sense). *A bag contains 3 pairs of blue socks, 4 pairs of white socks, and 5 pairs of black socks, with the individual socks loosely mixed together. How many socks need to be drawn from the bag blindly (that is, without looking at the socks before or after each draw) to ensure that we got a pair?*

Question 2 (requires basic arithmetic or algebra). *Three men, G, L, and T, decide to have a mini-party. At the agreed time, G shows up with 5 bottles of wine, and L with 3 bottles; T, who had worked beyond the closing time of the*

---

[1] American journalist and social critic, 1880–1956.
[2] American educator, 1858–1945.

liquor store, offers to pay his share in cash and is told by his friends to fork out $24. After the party, the money is divided between G and L (who had brought the wine) so that each of the three has contributed an equal share. How is the money split, if all the bottles of wine are identical and purchased from the same store?

"I suggest you give these questions to your students," JJ said.

"They are too easy. My students won't have any trouble with them."

"You've got a lot of faith in the public school system."

"Why shouldn't I? What's wrong with it?"

"I'm not convinced that it does the job it's supposed to do."

I wriggled on my chair to get into a comfortable position. "Can you be more specific?"

"Of late," JJ said, "the thinking and actions of the education authorities have been heavily influenced by certain 'modern' scholastic doctrines that claim to help all children have a fair crack at useful learning, but whose practical effect seems to be quite the opposite. If this path is not abandoned, the system might soon face a serious crisis, from which recovery will be slow and painful."

"We've heard such gloom-and-doom prophecies before. How do I know yours is not just scare-mongering?"

JJ rubbed his cheek. "I've already told you that one of the things our human-shaped team does is conduct population surveys. To make things simple, from now on, whenever I want to mention one of my colleagues, I'll use the generic name Gan, and when I refer to our interviewees, I'll call them JOE. Unspecified students will be called Stu. Frequently, the answers in the snippets you will see and hear come from different persons."

"And who was surveyed this time?"

"A large number of public school teachers."

"With the VOTSIT?"

"What else? We need to ensure that the subjects always tell the truth. Here are a few typical answers Gan received on this occasion. Don't be unnerved by what you'll see next: the VOTSIT plays back its recordings in a very special way."

As he said that, something strange happened, something almost impossible to describe. I felt as if the space around me had been cut out of the universe and transformed into a small cubicle occupied only by myself and what seemed to be a 3-D projection volume where people sat and talked. Their faces, words, and gestures were exceptionally clear. A kind of futuristic private viewing room, you might say. I watched and listened with fascination.

GAN: Why did you become a teacher?

JOE: I always dreamed of taking young minds, fresh and eager, and enriching them with knowledge and a desire for improvement. I wanted to play a part in fashioning the intellect and character of those who will replace us when we are no longer here.

GAN: Do you get any satisfaction from your job?

JOE: Not to the extent I expected. In my view, the public school system appears to be working—if that's the right word—like a communist economy: everything is bureaucratically planned, operational directives are issued from the top and must be followed, there is very little room for innovation and very little encouragement for efficiency at grassroots level, and productivity measures are designed to show results that are not really there. No wonder we hear so many complaints from parents and employers. Speaking as a parent, I would feel deeply betrayed to see my son and daughter handicapped by their second-rate education when they compete in the job market with properly schooled candidates.

GAN: How do the teachers survive in their jobs?

JOE: The seasoned ones make the best of a bad situation and eventually get something out of nothing. The less experienced are having a hard time of it and may be forced to adopt desperate tactics to keep their heads above water. This shouldn't come as a surprise. Not long ago, our state education department 'realigned' the end-of-instruction algebra exam to bring it closer, they explained, to what ninth-graders were being taught. The rate of demonstrated proficiency more than doubled! When the generals get away with fraud on that scale, why should the foot soldiers be condemned if they massage a few scores?

GAN: Are you contemplating a change of career?

JOE: Not yet. But the absence of incentives and rewards for teaching excellence in the public system is making me think very seriously about moving to the private sector, where the professional landscape is less dreary and quality has a better chance to prevail.

GAN: Can bad teachers be fired?

JOE: It depends what you mean by 'bad'. If you mean someone who has assaulted a student, then the answer is 'perhaps', but only at the end of a lengthy and tortuous procedure that may take several years, during which the person is suspended on full pay. If, on the other hand, you mean incompetent, then the answer is 'practically never'.

GAN: How is classroom discipline?

JOE: In the inner city school where I work, classroom discipline is a bit of a joke. Young people should venerate and look up to their teachers, as Seneca[3] said. But they do nothing of the sort. To make a class of students functional, democratic principles are not the answer. Someone has to lay down the law and the rest must follow, otherwise things quickly degenerate into chaos. Unfortunately, faced with lack of parental cooperation and a judiciary that seems to have lost its bearings when it comes to civil liberties, we don't dare say 'boo' to our charges for fear that they might sue us.

GAN: What about your unions?

JOE: When the labor force started organizing itself, unions played a very positive role. They formed a bulwark against exploitation of the workers by their bosses.

---

[3] Lucius Annaeus Seneca (Roman philosopher and dramatist, c4 BC – AD 65) actually said, *"Praeceptores suos adulescens veneratur et suspicit,"* because he spoke Latin.

Later, however, some of them veered toward a dubious kind of militancy, tending to lose sight of their original purpose. Are the teaching unions helpful to their members? I'm not sure. It's the unions who make it practically impossible for the public schools to get rid of the occasional bad apple and dictate that everyone be paid the same, regardless of individual performance and results. I often wonder how instructors who prefer to abandon their classroom duties to march and waive slogans at rallies can claim that they are fulfilling their contractual obligations. To me, they seem more intent on playing politics than doing their job. The students aren't the main concern of the unions. As a former president[4] of the American Federation of Teachers admitted, "When schoolchildren start paying union dues, that's when I'll start representing the interests of schoolchildren."

GAN: Would more money help?

JOE: Both the politicians and the unions firmly believe—albeit for different reasons—that the below-par quality of school education can be improved only if more funding is made available. This is one of the most blatant, yet widely accepted by the public, fallacies of our time. True, compared to other professionals we are poorly paid, but throwing money at the system will not by itself cure its ills. The extra resources get squandered on projects that are of no tangible benefit to us or our customers.

GAN: What, in your view, is fueling the decline of education?

JOE: Misunderstood and misapplied egalitarianism. Quite often, bad things are not caused by bad people, but by good people whose good intentions lead them to an unhappy place, where they turn obsessive and then tyrannical. Consider the following argument:

Every child must get a diploma on finishing school.
Not all children are equipped to become rocket scientists or brain surgeons.
All children can learn to flip burgers.

Therefore, every child can get a diploma if school education is pitched at burger-flipping level.

This logic works if you accept the premise, which is what those in charge appear to have in mind for public schools. So they aim for the lowest common denominator, dumbing down the curriculum and trivializing it to such an extent that the results shoot up to giddying heights across the board. The outcome of this 'universal success' strategy is crop after crop of semidocts, whose insufficient general knowledge is exposed, for example, in TV game shows, newspaper articles, Hollywood productions, and so on. If you were to turn an old quote on its head, you could say that, under the pretext of 'syllabus diversification', public schools are teaching less and less about more and more, until their students will know nothing about everything.

GAN: Can you give me a specific example?

JOE: Just before the end of last year, a couple of education hotshots visited our school and presented some new ideas they were planning to recommend for

---

[4] Albert Shanker, 1928–1997.

nationwide adoption. Among other things, they suggested that all testing should consist of just one comprehensive but brief final exam, which would determine the students' grades. Then they administered a specimen test to our seniors and discussed the annotated papers with us. Here is the test, a compilation of the most unusual answers (with the original spelling), and the grader's comments.

1. *Physics Question.* If a ball rolls off a table, would it fall down or go up?

   Student's Answer. Go up.

   Grader's Comment. Theoretically speaking, this could be construed as correct if the student observes the ball's movement from a handstand position. It is equivalent to directing the vertical axis downward.

   Verdict on Answer. Acceptable.

2. *Chemistry Question.* What is the most plentiful drinkable substance on Earth?

   Student's Answer. Cola.

   Grader's Comment. Obviously, the student drinks nothing but varieties of this product. His answer is based on direct experience, which takes precedence over objective fact. Besides, cola has a significant water component.

   Verdict on Answer. Acceptable.

3. *Biology Question.* Do you write with your fingers or your toes?

   Student's Answer. Both.

   Grader's Comment. Judging by the graphologic evidence on this test, the student must be right.

   Verdict on Answer. Acceptable.

4. *Mathematics Question.* If algebra could be heard, what would it sound like?

   Student's Answer. Hells bell's.

   Grader's Comment. The student seems to have less experience with this subject, which causes him anxiety. His response is, therefore, a true reflection of his frame of mind.

   Verdict on Answer. Acceptable.

5. *History Question.* What person, from the past or present, do you respect and admire most, and why?

   Student's Answer. Look Skywakka. Man, he distructed the dark side real good didn he.

   Grader's Comment. A positive choice of hero, stimulating and inspirational.

   Verdict on Answer. Acceptable.

6. *Geography Question.* What is the name of the place where you have lived most of your life so far?

   Student's Answer. The hi street mall.

Grader's Comment. Within the student's interpretation of the question, a candidly truthful admission.

Verdict on Answer. Acceptable.

7. *Literature Question.* Who is your favorite writer, and why?

   Student's Answer. My cuzzin Jim. You shuld see what he doz wiv a can of spray paint.

   Grader's Comment. A writer means someone who writes, and cousin Jim certainly comes into that category. The student also shows appreciation for the graphical artistry of Jim's writing.

   Verdict on Answer. Acceptable.

8. *Grammar Question.* Compose a two-word sentence.

   Student's Answer. Dunno any.

   Grader's Comment. Technically, this is a sentence fragment, but the missing subject "I" is so obvious that the sentence may be regarded as virtually complete.

   Verdict on Answer. Acceptable.

9. *Vocabulary Question.* What does the word "doe" mean?

   Student's Answer. Money.

   Grader's Comment. Since the question does not intend to test the respondent's spelling skills, we must be guided by the sound of the word, which justifies the chosen synonym.

   Verdict on Answer. Acceptable.

10. *Social Science Question.* Who is known as the "Commander in Chief"?

    Student's Answer. My gang leader.

    Grader's Comment. The subject has reacted to his own immediate life environment. He has, in fact, understood that the question refers to a figure of authority.

    Verdict on Answer. Acceptable.

Ten out of ten, and the end of failure as we know it. I rest my case.

GAN: How does the country continue to boast so many top scientists, humanists, and artists if its education system is as inadequate as you say?

JOE: Homegrown prominent people usually come from the better schools. The few educated at run-of-the-mill establishments who make it to the top do so under their own steam. For hard as you may try, you cannot stifle excellence. Talent and resolve will eventually slash through any constraining wrapper and rise to the surface. But think how dramatically the number of distinguished professionals would grow if the public school system was solid and sound!

GAN: What should the government do?

JOE: Take a close look at the so-called education experts to see if they speak from knowledge based on classroom experience or are just spinning airy-fairy theories. Ask them to try out these theories by enrolling *their* children—and teaching for a few months—in a school like mine. Keep politics out of education. Persuade the unions to adopt a more constructive attitude. Raise teachers' salaries to a decent level and introduce pay differentials based on a set of objective and transparent criteria. Find an acceptable way to maintain discipline in the classroom. Banish the one-size-fits-all mentality: it's counterproductive. Place the high-flyers on a fast track and allow them to advance at their own pace; they are the movers and shakers of tomorrow, whose unimpeded growth will benefit the entire society. Expand vocational education. Admit that not every child is cut out for college and make it clear that, to function properly, the country needs in equal measure good doctors, plumbers, accountants, hairdressers, engineers, and office secretaries. Let school resemble real life, where substandard work always brings about failure.

GAN: Will this happen?

JOE: I don't know. At the moment, the authorities are reluctant to acknowledge that government-run education is mediocre and pussyfoot around the issue instead of actively seeking to restore learning's appeal and usefulness.

"Pretty radical stuff," I said when JJ's recording ended. "Are public schools that bad?"

"You heard it from the horse's mouth: disgruntled teachers, overworked and underpaid, who do what they can in a system long on rhetoric and short on achievements. One of your former presidents[5] said, 'The progress of the country as a nation can be no swifter than its progress in education.' How very true. This is why the MICQ of the education gurus and their followers is relegated to the 30–34 band. The BRITEs should close ranks and find a solution quickly if your country is to maintain and consolidate its position in the world's top echelon."

"What's a semidoct?[6] I haven't heard this word before."

"It's an import. In some Latin-based languages it designates one who has little and superficial culture, but believes oneself to be cultured. A pseudo-intellectual. The kind of person Alexander Pope[7] had in mind when he wrote that 'a little learning is a dangerous thing'." JJ took a deep breath. "Now what about some mathematically flavored jokes? You promised."

"I did nothing of the sort. However, I happen to remember three little stories that might be suitable for your collection."

"Well, I'm ready. Fire away."

"The first is in the form of a mathematical assertion and sounds something like this."

Theorem. The less one knows, the more one gets paid.

---

[5] John Fitzgerald Kennedy, 1917–1963.

[6] From the Latin *semidoctus:* half-taught.

[7] English poet, 1688–1744.

Proof. Generally accepted wisdom has it that

(i) knowledge is power;[8]

(ii) time is money.

Also, according to elementary physics,

$$\text{power} = \frac{\text{work}}{\text{time}}.$$

By assertions (i) and (ii), this equality can be rewritten in the form

$$\text{knowledge} = \frac{\text{work}}{\text{money}},$$

which yields

$$\text{money} = \frac{\text{work}}{\text{knowledge}}.$$

It is thus clear that for the same amount of work done, as knowledge decreases, money increases. Furthermore, since, for any amount of work, money grows without bounds as knowledge dwindles to zero, we deduce that a completely ignorant person can amass unlimited wealth with hardly any effort. This would explain why many well-paid positions of authority are taken by incompetents.

JJ tapped his fingers on the table. "So who needs education, eh?"

"And here's the second one," I said.

Just as the bell rings at the end of his class, the history-of-religion teacher asks little Ben, "Tell me, who brought down the wall of Jericho?"[9]

Ben goes pale and, in a shaky voice, says, "It wasn't me, sir, I swear. I didn't go anywhere near it."

The teacher is appalled. "I don't believe this!" he exclaims, and storms out of the classroom.

Ben walks home after school and tells his parents what the teacher had said. "You're sure you never touched it?" his father asks him sternly, knowing that Ben is not exactly a model boy.

"Very sure," Ben says. "Tim, Marty, and I spent all the breaks together, and we didn't mess with no stupid wall."

Next day, Ben's father goes to see the principal, who is in his office, talking to the bursar. "My son never lies," the father points out. "If he says he didn't do it, then he didn't do it. I don't like to see him wrongly accused."

---

[8] Francis Bacon (English philosopher, statesman, and essayist, 1561–1626) said it first: *Ipsa scientia potestas est.* (Latin)

[9] An ancient city in the land of Canaan, conquered and destroyed by the Israelites at the end of the 15th century BC.

"Have no fear, sir," the principal assures him. "I will personally look into this matter."

When Ben's father leaves, the principal turns to the bursar and shakes his head. "You hear something funny every day, don't you?" he says. To which the bursar replies, "Why make such a fuss over a kiddie's prank? Let's get a couple of estimates, repair the wall, and put this whole thing behind us."

"The state-school education shining through," JJ observed caustically. "It brings to mind Will Rogers's[10] quip that 'the schools ain't what they used to be, and never was'. And the third story?"

"It has to do with one of the very lucrative nonacademic occupations you mentioned earlier."

A professor of mathematics notices that his kitchen sink has sprung a leak, so he calls a plumber. When the man arrives, he crouches under the sink, replaces a gasket and, in a couple of minutes, has the sink perfectly operational again.

"That will be 100 dollars, please."

"What?? A hundred bucks for replacing one lousy gasket?" the mathematician says indignantly. "I'm a professor and don't make anywhere near that much money for two minutes' work."

"Well, sir, then perhaps you should consider changing your profession and becoming a plumber. It's a good, steady job, the pay and hours are great, and our company certainly needs more people. But if you apply for a position, don't tell them that you have any education beyond the seventh grade. They reject overqualified candidates out of hand."

The professor thinks about that and resolves to give it a try, so he joins the company and becomes a plumber. Right away, his income grows dramatically and the quality of his life skyrockets.

A year later, the company's CEO decides that the workers must improve their general education and enrolls them all in night-school eighth grade. The professor joins the other plumbers and goes to the first class, which happens to be mathematics.

"I need to find out what you know before I can start you on new material," the teacher says. "You, there in the front row," and she points to the mathematician, "what is the formula for the area of a circle?"

The professor, who has not used that formula for a long time, has by now forgotten it, so he goes to the board and starts sketching, writing functions, differentiating and integrating, and in the end comes up with $-\pi r^2$. He knows that something is wrong, because an area cannot be negative, so he starts over again, and again, and again, but every time he arrives at the same result. Thoroughly frustrated, he turns to the class and looks at his fellow plumbers, who are all whispering frantically, "Swap the limits of integration!"[11]

---

[10] William Penn Adair Rogers, American comedian, humorist, and social commentator, 1879–1935.

[11] For the uninitiated: swapping the limits of integration changes the sign of the result.

I thought I saw a shadow of sadness wash over JJ's face. "It's the same on all primitive planets: mathematicians are never rewarded fairly."

I wanted to take issue with his use of the word 'primitive', but I didn't get the chance. "Thanks for the stories," he said. "I've really enjoyed our little chat. If all goes well, we'll talk some more next time." Then, unceremoniously, he sprang to his feet and promptly walked out of the bar.

Notes After the Meeting

Here are the answers to JJ's questions together with some brief personal comments.

Solution to Question 1. To get a pair, you need to draw two socks, of course, because a pair, in the strict interpretation of the word, consists of two elements.

A colleague who tried this question on a class of freshmen summed up her students' answers in the following table.

Percentage	Answer	Verdict
10	2	correct
80	4	incorrect (see comment (i))
5	11 (for a blue pair) 10 (for a white pair) 9 (for a black pair)	incorrect (see comment (ii))
4	20 (for a blue pair) 18 (for a white pair) 16 (for a black pair)	incorrect (see comment (iii))
1	full analysis	excellent (see comment (iv))

(i) These students must have thought that the question meant how many socks need to be drawn to have a pair *of the same color*. For *that* question, their answer would be correct because, by the pigeonhole principle,[12] the worst case scenario is that the first three socks drawn from the bag are of different colors, so the fourth will necessarily match one of the first three.

(ii) The answers in this group try to detail how many socks have to be drawn to get a pair of each particular color, and, again by the pigeonhole principle, they would be correct if 3, 4, and 5 were the numbers of *individual* socks of each color.

---

[12] Also known as Dirichlet's box principle: if n pigeons are placed in $m < n$ pigeonholes, then there is at least one hole with more than one pigeon in it.

Johann Peter Gustav Lejeune Dirichlet: German mathematician, 1805–1859.

(iii) Comment (ii) is still valid, but these students worked out correctly that the numbers of socks of each color are, respectively, 6, 8, and 10. For example, if we want to make sure that we have a blue pair, the worst case scenario says that we do not draw any blue sock at all until we have drawn all the white and black ones first, 18 in all; then the next two will be blue. (A single blue sock could be drawn earlier, but that doesn't change the answer.)

(iv) Only one person answered correctly the original question *and* gave the alternative interpretations (i) and (iii), the latter after saying "but if the question really means...."

The respondents covered by comments (i)–(iii) were victims of a false sense of expectation. The question builds up a complex background of three different colors and numbers that, they think, must imply something deeper than the boring obvious. Certainly. It implies a bit of mischief. If they had doubts, they should have done as in comment (iv).

Students should remember to read a question well before answering it. They must not make up their own questions. In an exam this could prove very costly.

Solution to Question 2 (arithmetic). *Since each man contributed an equal share and T paid $24, it means that each of the three equal shares was $24, so the cost of the party (8 bottles of wine in total) was*

$$\$24 \times 3 = \$72.$$

Therefore, the price of a bottle was

$$\$72 \div 8 = \$9.$$

G had spent
$$\$9 \times 5 = \$45,$$

so, since his share was only $24, he got back

$$\$45 - \$24 = \$21$$

from the $24 paid by T.

L had spent
$$\$3 \times 9 = \$27,$$

so he got back

$$\$27 - \$24 = \$3.$$

Solution to Question 2 (algebraic). *In dollars, let x be the price of a bottle of wine, and let y_1 and y_2 be the amounts of money that G and L got back, respectively.*

G spent $5x$ on the wine and got back y_1, so his net contribution to the party was $5x - y_1$.

L spent $3x$ on the wine and got back y_2, so his net contribution was $3x - y_2$. T's contribution was $y_1 + y_2 = 24$.

Since all contributions were equal, we can write

$$5x - y_1 = 3x - y_2 = 24,$$

which is equivalent to the system of simultaneous equations

$$\begin{aligned} 5x - y_1 &= 24, \\ 3x - y_2 &= 24. \end{aligned} \qquad (3.1)$$

If we add these equations side by side, we arrive at

$$\begin{aligned} & 8x - (y_1 + y_2) = 48 \\ \Leftrightarrow\ & 8x - 24 = 48 \\ \Leftrightarrow\ & 8x = 72, \end{aligned}$$

from which we deduce that

$$x = 9. \qquad (3.2)$$

Then, by (3.1) and (3.2),

$$\begin{aligned} y_1 &= 5x - 24 = 5 \times 9 - 24 = 21, \\ y_2 &= 3x - 24 = 3 \times 9 - 24 = 3. \end{aligned}$$

For those who don't scare easily I suggest a refined version of the question, where G brings 3 bottles more than L, the price per bottle (a positive integer) paid by L is $2 more than the price per bottle paid by G, and T is told to pay $32. All I will say as a hint is that this version has two distinct answers. Try and find them both.

The first solution to Question 2 shows that one should not dismiss lightly the magnificent simplicity and elegance of arithmetic. Bludgeoning through with an algebraic argument is not always the most satisfying way to solve a problem. However, make the problem more general, as mentioned above, and algebra will quickly come into its own.

A word of caution: to avoid a run-in with the law, underage readers should replace 'wine' with their preferred soda. It won't be much of a party, but the mathematics remains the same.

A Word to the Wise

Equality and Implication

Propositional logic includes a couple of useful symbols that students sometimes misuse. Let α and β be two mathematical statements. The symbolic relationship

$$\alpha \Rightarrow \beta$$

means 'α implies β'. The symbolic relationship

$$\alpha \Leftrightarrow \beta$$

means 'α is equivalent to β' (that is, α implies β and β implies α).

☺ DO

Here is an example of clear and correct layout:

$$\begin{aligned} & a+b=0 \\ \Rightarrow\ & c(a+b)=0 \\ \Leftrightarrow\ & ca+cb=0. \end{aligned}$$

The equality on line 1 implies that on line 2, because line 2 is obtained by multiplying line 1 by c. But line 2 does not imply line 1: line 2 implies either line 1, or $c=0$. Hence, lines 1 and 2 are not equivalent.

Lines 2 and 3, on the other hand, are equivalent. Line 3 is obtained from line 2 by multiplying c into the two terms a and b; line 2 is obtained from line 3 by factoring out c.

☠ DON'T

As already explained, the following is wrong:

$$\begin{aligned} & a+b=0 \\ ☠\ \Leftrightarrow\ & c(a+b)=0. \end{aligned}$$

☺ DO

The explicit solution of an equation should be laid out like this:

$$\begin{aligned} & x^2-9=0 \\ \Leftrightarrow\ & (x+3)(x-3)=0 \\ \Leftrightarrow\ & x+3=0 \quad \text{or} \quad x-3=0 \\ \Leftrightarrow\ & x=-3 \quad \text{or} \quad x=3. \end{aligned}$$

☠ DON'T

Many times, work is written up negligently. Consider the following sample:

$$\begin{aligned} & x^2-9=0 \\ ☠\ \Rightarrow\ & (x+3)(x-3) \\ ☠\ \Rightarrow\ & x=3. \end{aligned}$$

This is wrong on three counts:

(i) Line 2 is not an equation any more; it has lost '$= 0$' at the end.

(ii) The (corrected) formula on line 2 should be preceded by the equivalence sign $\Leftrightarrow$.

(iii) The (corrected) line 2 does not imply line 3 because $x = -3$ is another possible consequence of line 2.

Some other times the equality sign is wrongly used in place of an implication sign. Thus, one may see

$$☠ \quad 2x = 2y = x = y$$

instead of the correct form

$$2x = 2y \quad \Leftrightarrow \quad x = y.$$

SCAM 4

The School Mathematical Education

> *What would life be without arithmetic, but a scene of horrors?*
> Sydney Smith[1]

> *Arithmétique! Algèbre! Géometrie! Trinité grandiose! Triangle lumineux! Celui qui ne vous a pas connues est un insensé!*[2]
> Comte de Lautréamont[3]

I met JJ again at a mathematics conference sponsored by Duke University in Durham, North Carolina. Just like the first time, he was sitting alone at a table in a corner of the hotel bar, gazing at me across the room with dark, inscrutable eyes.

As soon as I sat down, he pushed a piece of paper in my direction. "Here," he said laconically. "For your students."

I glanced at the paper and smiled. The text read something like this.

Question 1 (requires basic arithmetic). *An old cowboy dies and his three sons are called before an attorney in town for the reading of their father's will. After some legalistic jargon, they hear the following important passage:*

"All I have in this world I leave to my three sons, and all I have is just a few horses. To my oldest son, who has been a great help to me and done a lot of hard work, I bequeath half of my horses. To my second son, who has been almost as helpful but, being younger, has worked a little less, I bequeath a third of my horses. My youngest son likes drinking and womanizing and hasn't helped me any. However, he is still my son and I cannot let him go empty-handed, so to him I bequeath a ninth of my horses. And this is my last will and testament."

The sons go back to the corral and count the horses, wanting to divide them according to their pa's exact wishes. But they run into trouble right away when they see that there are 17 horses in all and that they cannot do a proper division. The oldest son, who is entitled to half—that is, $8\frac{1}{2}$ horses—wants to take 9. His brothers immediately protest and say that he cannot take more

---

[1] English writer and clergyman, 1771–1845.

[2] Arithmetic! Algebra! Geometry! Grandiose trinity! Luminous triangle! Whoever has not known you is a fool! (French)

[3] The pen name of Isidore Lucien Ducasse, a Uruguay-born French poet, 1846–1870.

than what he is entitled to, and if this means that they have to butcher a horse to do a fair deal, then so be it....

Just as they are ready to start a fistfight, a stranger comes riding by and stops to ask what the commotion is all about. Learning of their predicament, he dismounts, lets his horse mingle with the others, and tells the three brothers that he can solve their problem.

"There are 18 horses in the corral now," he points, "which is a much better number for the kind of counting you need to do. Your share is a half," he says to the oldest brother, "that is, $18 \div 2 = 9$ horses. Yours is a third," he says to the second brother, "in other words, $18 \div 3 = 6$ horses. And yours is a ninth," he tells the youngest brother, "which comes to $18 \div 9 = 2$ horses. Now $9 + 6 + 2 = 17$ horses, exactly what your father left you. Go and get your shares." And the stranger mounts his own horse (the eighteenth one) and rides quietly into the sunset.

How was this kind of division possible?

Question 2 (requires elementary logic). Fred has fallen in love with a beautiful princess. When the young lady indicates that she feels the same about him, Fred asks her father, the king, for his daughter's hand in marriage. But the king is not amused and tells Fred that he is guilty of lèse-majesté[4] and must be punished severely. Fred is taken to a long hall with one door at each end and an armed guard in the middle. "You think you are so smart," the king scoffs, "that you can rise above your humble commoner status by marrying my daughter. Let's see if you are smart enough to save your life." The king points to the two opposite doors. "One of these doors leads to certain death, the other one to freedom. The guard is from Funnyland, a country where every citizen belongs to one of two distinct tribes: the Truth Tellers (TTs), who always tell the truth, and the Pathological Liars (PLs), who always lie. Given that the guard knows what's behind each door, you are allowed to ask him a single question, phrased in such a way that he can answer it silently. After that, you pick a door, exit, and meet your fate. You've got two minutes."

Fred does as instructed and, a few moments later, walks away to safety and freedom. What was his question, and which door did he choose?

"Interesting," I said. "Thanks." I folded the paper and put it in my pocket. "You think my students won't be able to come up with the correct explanation?"

"Some might, some might not. It depends on how much logical thinking they learned from their school mathematics teachers."

"There are some pretty good math teachers out there."

"There are some *excellent* math teachers out there," JJ replied. "People may not be aware of this, but the impact school math teachers have on their students is enormous and tends to last a lifetime."

---

[4] Offense against the sovereign authority. Literally: hurt majesty. (French)

"In which way?" I asked, as if I didn't know....

"The students taught math by good teachers will always think logically, construct arguments that have a hypothesis, a body of reasoning, and a conclusion, and obey the ten mathematical commandments, even though they may not remember much of the technicalities of algebra and calculus. Those whose teachers have a more relaxed attitude take forward with them a harmful assortment of 'principles' that will spell trouble for their careers."

"What kind of principles are you talking about?"

"Here is a small sample," JJ said. "I'm sure you've seen them applied at some time or other."

The Principle of Least Headache. If a can be equal to b, it is.

The Principle of Unlimited Confidence. If a is not equal to b, it ought to be.

The Principle of Convenient Choice. If several mutually exclusive conclusions can be drawn from the same premise, always choose the one that suits your argument.

The Principle of Wishful Thinking. I know that what I write is gibberish, but the teacher might still find something worth an extra point in it.

"Can you guess what's damaging your young generation most?" JJ asked.

"Fast food?" I chanced. "Video games? Parental neglect?"

"I meant the kids' mathematical education."

"I'm sure you'll tell me."

"It's the theories spun by some influential educationists. I already mentioned them at our last meeting."

"Yes, but you never explained why."

"An example is better than a lengthy explanation. My colleague Gan spoke once with a professor of policy studies in mathematical education, Pol, who was preparing a research proposal for external funding and wanted Gan's opinion on it. Their conversation, recorded by Gan on his VOTSIT, went something like this."

GAN: The project claims to recommend a course of action that will make mathematics more accessible to disadvantaged social groups. How do you suggest this could be achieved?

POL: We must simplify the mathematics taught in the classroom. Arithmetic and algebra will be cut out of the syllabus and replaced by calculators. No proofs will be shown, ever. Rigor will be considered a sign of weakness. The students will be directed to see mathematics as an element of creative art rather than a means of acquiring good computational skills. Group learning will be the only method of work and the only basis for assessment. Since all schoolchildren are the same, if one is worth a certain grade, then all of them must get that grade.

GAN: But, I'm told, this is already being done in the public schools. What's *novel* in your proposal?

POL: First, there was Math. Then came New Math. Now we have Whole Math. My model, called New Whole Math, argues for the reform to be pushed all the way.

For example, students doing sums in their head, learning individually, or getting correct answers should be disciplined. Failure should formally be eliminated from judging standards and results. And teachers and parents who criticize the program should be declared enemies of the state and thrown in jail.

GAN: The title baffles me: "A Self-Sustaining Strategy for Disseminating Mathematical Skills in Socially Challenged Strata Through Higher-Order Thinking, Teacher Enhancement, and Community Engagement Models." Can you say that in plain English?

POL: I can, but I won't. It wouldn't sound impressive enough. The whole point is not to let the members of the review panel understand what it means. I'd get no money if they did.

GAN: How much money are we talking about?

POL: Five million bucks over three years.

GAN: Five million?! How can you justify that kind of expenditure?

POL: Easily. My group consists of three investigators. First, we pay our own salaries for the summer months, which, as you know, are over and above an academic's nine-month contract. Then we need to attend 14 professional conferences: in Honolulu, Acapulco, Melbourne, Las Vegas, Rome, Vienna, Rio de Janeiro, Hong Kong, Tokyo, Stockholm, Capetown, Nice, St. Moritz, and Timbuktu.

GAN: Why?

POL: To communicate our findings to the world. And to hear what other colleagues in the field are doing elsewhere.

GAN: Yes, but *fourteen*?

POL: That's just a conservative estimate. Things in our line of business move very fast. Anyway, we also need to buy new desk computers, laptops, and two grand pianos.

GAN: Why laptops?

POL (indignantly): How can you say that? You know research. You don't tell a great idea when to come to your mind. You may get one while you are having breakfast with the family at home.

GAN: Or while you are attending those 14 conferences.

POL: Precisely.

GAN: And the pianos?!

POL: All work and no play is detrimental to our well-being. We need an outlet for our mental stress. Then we need to consult several experts.

GAN: But you *are* the experts.

POL: Expertise is relative. There are always people more expert than you. Then we hire two full-time assistants.

GAN: To do what?

POL: Collect the data and perform the necessary analyses. They will also play the pianos.

GAN: If they do all that, then what will *you* do?

POL: Interpret the results and make recommendations.

GAN: To whom?

POL: To whoever's interested.

GAN: What if nobody's interested?

POL: Oh, there'll be plenty of receptive eyes and ears. On the one hand, you have the employees of local authority education departments with consistently poor results in their schools. In desperation, they will try anything to turn things around. On the other hand, there are the functionaries of education departments with good results, who, wanting to persuade their paymasters that they are progressive and forward-thinking, will eagerly jump on the bandwagon of trendy new ideas.

GAN: Won't they risk going from good results to bad?

POL (with disarming frankness): Not our problem, once we have secured the five million.

GAN: This is cynical. You are squandering taxpayers' money.

POL: It's how the game is played. There are funds to be spent by the government each year on educational research. If we don't get a piece of them, someone else will. So it might as well be us.

GAN: What do you think of reformed calculus?

POL: Best idea of our times. I wish I had it.

GAN: Just as I expected: your MICQ is no higher than 34.

"Well?" I said. "Did Gan tell Pol what he thought of his proposal?"

"Yes."

"And?"

"He still submitted it. Unchanged."

"When?"

"Year before last," JJ said.

"So what's Pol doing now? Rewriting the old text for resubmission? Brewing some new initiatives?"

"No. He's busy spending his five million dollars. People like him can be very persuasive when they deal with the uninitiated."

"How would you use the money if it were given to you?"

"In a logical way. I'd gather the schoolteachers together periodically for refresher courses in basic mathematics from people who know what the subject is about and how it should be taught. That might make a difference." JJ leaned back in his chair. "Now," he said, "it's time for your stories."

I finished the last of the coffee and pushed the empty cup aside. "OK, here we go. Will you take three again?"

"With a mathematical dimension?"

"Judge for yourself. First one coming up."

Little Jimmy's parents, staunch supporters of a secular upbringing and education for their son, are so dismayed by his performance in mathematics at the state school where he goes that they decide to relax their principles and send him to a private school in town, which is run by a religious order of priests and has an excellent reputation.

From the very first day at the new school, Jimmy's attitude and working habits change dramatically. As soon as he arrives home, he says 'hello' to mum and dad, goes up to his room and studies hard, and does not come down again until dinner.

After dinner, he declines to watch TV and goes back to his room for more studying. This happens every day without exception, week after week after week. When he brings home the first report card, his parents are astonished to see that Jimmy got an A in mathematics.

"What made the difference?" they ask him. "Are the teachers better at this school?"

"Not really," Jimmy says.

"Is the library roomier and quieter? Are the textbooks easier to read?"

"They are more or less like the other ones."

"Did you suddenly develop a taste for working with numbers and now you understand how to handle them?"

"I hate numbers."

The parents are mystified. "What, then, made you do so well?"

"On the first day," Jimmy explains, "they took us to a large hall, where I saw this little guy nailed to a 'plus' sign. That told me it was time to quit fooling around and get down to some proper learning."

"If only it were that simple..." JJ mused.

"In view of what Gan discussed with Pol, the second story might please you," I said.

A man with an IQ of 180 dies and arrives at the Pearly Gates, where the Gatekeeper tells him that the policy in Heaven is to mix people of different intelligence levels in the dormitories in order to avoid accusations of elitism. He then orders a winged escort to take the man to his eternity quarters.

"You'll have three roommates," the escort says as they enter the Elysian apartment. "One of them has an IQ of 150."

"This means that I'll be able to talk higher mathematics with him."

"The second one has an IQ of 120."

"We'll be discussing particle physics a lot."

"And the third one," the escort says, "has an IQ of 90."

The man perks up. "Great! I've always wanted to ask an educationist how his kind managed to emasculate the public school system."

"You altered this one for me, didn't you?" JJ's eyes seemed to sketch a smile. "The original couldn't have had an educationist in it."

"Why not?"

"An IQ of 90 would be too low for him. Educationists are very clever people. You need a reasonably high degree of intelligence to extract money from government agencies on false pretenses and make them feel that you've done them a favor."

"But Gan said that Pol's MICQ was no higher than 34."

"Have you forgotten already? The MICQ and your IQ are not the same thing. Brains alone do not secure a respectable MICQ score for a person. Pol supports reformed calculus, which is an infringement of the Seventh Mathematical Commandment and identifies him as an unBRITE. Anyway, I appreciate your gesture. Will you tell me now what type of person the original version of the story referred to?"

"No. There are certain topics that I won't discuss. But here's the last of today's installments."

The teaching of school mathematics has undergone substantial changes in the past few decades. Are these changes good or bad? Instead of a direct answer, let's consider an illustrative example: the formulation of a simple problem of arithmetic at ten-year intervals starting from the middle of the twentieth century.

1950s A farmer sells his produce for $100. His labor cost is 4/5 of the price. What is his profit?

1960s A farmer sells his produce for $100. His labor cost is 4/5 of the price, or $80. Is his profit (a) $0, (b) $100, or (c) $20?

1970s A farmer sells his produce for a set D of dollars. The cardinality of D is 100. Represent this by drawing 100 dots. The farmer's labor cost is a subset C of D, which contains 20 points fewer than D. When the points of C are eliminated from D, we are left with the set P of profit. What is the cardinality of P?

1980s A farmer sells his produce for $100. His labor cost is $80 and his profit is $20. Use the FARMER program on your calculator to verify that these figures match. Do you find math difficult?

1990s A farmer makes $20 by selling his produce. To do so, he drives his pickup truck to the market, poisoning the air with fumes and destroying the world's natural resources. How do you feel about that? Should the government take any steps against the farmer? All answers will get an 'A'.

2000s A farmer sells his produce for $100. His labor cost is $120. Use a creative accounting technique to show that his profit is $50.

2010s A fRmR tayX h/ Vges 2 da mRkit n 1ts 2 keep all da $$$ 4 himself. WDYT shd B dun 2 dis $_$ d00d? B kewl n cr8iv.[5]

"Droll, but closer to home than you may imagine." JJ put his hands on the table. "Thanks for the stories," he said. "I've really enjoyed our little chat. If all goes well, we'll talk some more next time." Then, unceremoniously, he sprang to his feet and promptly walked out of the bar.

Notes After the Meeting

Here are the answers to JJ's questions together with some brief personal comments.

Solution to Question 1. The explanation is trivial: all the fractions into which the whole is split must add up to the whole. The fractions listed by the

---

[5] Translation for those unfamiliar with cell phone texting: *A farmer takes his vegetables to the market and wants to keep all the money for himself. What do you think should be done to this greedy dude? Be cool and creative.*

old cowboy in his will do not do that:

$$\frac{1}{2} + \frac{1}{3} + \frac{1}{9} = \frac{17}{18} < 1.$$

It is now obvious why the stranger's horse helped solve the problem: it completed the whole ($18/18 = 1$) and was then left behind after the old man's 17 horses had been divided between the three brothers according to his instructions. Had the will listed the fractions as 1/2, 1/3, and 'the rest', the brothers, in order to carry out their pa's wishes to the letter, would have had no choice but to call in a butcher.

The three sons must have been happy because their father's wishes had been observed and they each got more than what they were entitled to. The stranger was surely happy because he had pleased the young men and had prevented an ugly and unnecessary fight. The only unhappy character in this story must have been the attorney. When he had drafted the old man's will, he may have spotted the error but kept quiet because he had expected to be hired to litigate on behalf of one of the disgruntled brothers. Then again, he may not have, for when it comes to figures, attorneys recognize only obscenely large ones.

Solution to Question 2. Fred could ask, for example, "If you belonged to the other tribe, which door would you point out to me as being the way to safety?" If the guard is a TT (Truth Teller), he knows that a PL (Pathological Liar) would point to the wrong door, so, because he himself tells the truth, he also points to the wrong door, truthfully indicating the other man's answer. If he is a PL, he knows that a TT would point to the correct door, but, himself being a liar, he points to the wrong door. Hence, whichever door is indicated by the guard, Fred exits through the opposite one.

The solution has a whiff of elementary algebra about it. If we think of truth as positive and falsehood as negative, then Fred's question forced a concatenation of statements of opposite 'signs' whose 'product' is always negative, regardless of the order of the individual positive and negative statements in the chain. This is why Fred chose the other door and not the one identified by the Funnylander.

Of course, the solution is not unique. Fred could have asked an alternative question, for example, "If another member of your tribe stood here, to which door would he direct me?" A TT guard would know that any other TT would tell the truth and point to the correct door, so he would also point to the correct door. A PL guard, on the other hand, would be aware that any PL would point to the wrong door, so, because he lies, he would point to the door opposite, which would again be the correct door. This can once more be explained by the algebraic analogy in which the product of two positives and that of two negatives are both positive. If he had asked this question, Fred would have exited through the indicated door.

As regards the lovebirds' predicament, rumor has it that Fred and the princess eloped to Gretna Green,[6] got married, lived a life of privation on welfare, then, five years later, divorced in Las Vegas. Which goes to show, at least for the young woman, that there are situations when father knows best.

A Word to the Wise

Exact and Approximate Equalities

Many people seem unaware of the difference between the mathematical meaning of *exact* and *approximate* when writing equalities.

 DO

If a chain of equalities contains at least one approximation, then the 'ultimate' equality obtained by transitivity is also approximate; thus,

$$\sqrt{2} + \sqrt{3} \cong 1.41 + 1.73 = 3.14$$
$$\Rightarrow \quad \sqrt{2} + \sqrt{3} \cong 3.14.$$

Notice the use of the sign '=' above. This is justified by the fact that, although 1.41 and 1.73 are approximations of $\sqrt{2}$ and $\sqrt{3}$, respectively, 3.14 is the exact sum of 1.41 and 1.73.

If we are asked to find the largest root of the abstract equation

$$x^4 - 10x^2 + 1 = 0,$$

then we should give the exact result

$$x = \sqrt{2} + \sqrt{3}.$$

But if we are told that we need to cut out a square sheet of metal with side equal to x meters, where x is the largest root of this equation, then we should give the approximate result

$$x \cong 3.14$$

because x is now a measurement that will be used in a practical situation.

 DON'T

A common error is to present the rounded-off result obtained with a calculating device as exact. For example, some write

$$\text{} \quad \ln 5 = 1.6.$$

---

[6] A small village in the southwest of Scotland, famous as a venue for marriages between persons who, owing to various restrictive legal or social practices, cannot get married in their home jurisdiction. The Scots themselves do not have such problems.

Others think that if they insert a lot of decimals, they are entitled to use the exact equality sign:

$$\text{☠} \quad \ln 5 = 1.6094379124341003746007593332262.$$

A question such as "What is the area of a circle of radius 2?" is almost always answered with

$$\text{☠} \quad \pi \times (\text{radius})^2 = 3.14 \times 4 = 12.56.$$

This is obviously wrong because $\pi \cong 3.14$ is an approximation, which also makes 12.56 an approximation. The exact value of the area should be given as 4π.

SCAM 5

Language, Grammar, and Punctuation

> *Even if you do learn to speak correct English, whom are you going to speak it to?*
> Clarence Darrow[1]

> *Die Grenzen meiner Sprache bedeuten die Grenzen meiner Welt.*[2]
> Ludwig Wittgenstein[3]

I met JJ again at a conference sponsored by Emory University in Atlanta, Georgia. Just like the first time, he was sitting alone at a table in a corner of the hotel bar, gazing at me across the room with dark, inscrutable eyes.

As soon as I sat down, he pushed a piece of paper in my direction. "Here," he said laconically. "For your students."

I glanced at the paper and smiled. The text read something like this.

Question 1 (requires basic arithmetic). *A railway track 500 km long forms a straight line between two points X and Y. At X, a locomotive, capable of a maximum speed of 150 km/h, is waiting, facing in the direction of Y. At Y, a second locomotive, capable of a top speed of 100 km/h, is also waiting, facing in the direction of X. A fly, capable of a top speed of 200 km/h, is resting on the front end of the locomotive stationed at X. When a signal is given, the two locomotives start hurtling toward each other at their maximum speed. At the same time, the fly shoots at its maximum speed, in a straight horizontal line, toward the locomotive coming from Y. As soon as the fly touches this locomotive, it turns around instantaneously and heads toward the locomotive coming from X, then again turns back toward the locomotive from Y, and so on, until the two locomotives crash, at which moment the fly is no more. What is the total distance traveled by the fly before it expired?*

Question 2 (requires basic algebra). *Can an irrational power of an irrational number be rational?* (See the definition of an irrational number on p. 54.)

"Interesting," I said. "Thanks." I folded the paper and put it in my pocket, waiting for JJ to broach, as usual, a subject of his choice. He needed no special invitation.

---

[1] American lawyer, 1857–1938; best known for defending John Scopes in the so-called 'monkey trial'.
[2] The limits of my language mean the limits of my world. (German)
[3] Austrian philosopher, 1889–1951.

"Whaddup bro?" he asked abruptly. "Lunch gave you stomach cramps?"

"No, I'm just tired from too many Hilbert[4] spaces. Let me guess: you've got language on your mind today."

"I do."

"Why language?"

"Because language is crucial to civilization. And because I have seen and heard it being mistreated too many times. Since you, humans, are not telepathic, language is practically your only means of communication. You should care for it and treat it well, not abuse it. Just the other day, my colleague Gan related to me a conversation he'd had with a professor of linguistics, Lin. Watch his VOTSIT recording."

GAN: Do you think that the recent trend in spoken English to sprout meaningless slang is a positive development?

LIN: I totally do. What does it matter if we use literary English or slang? The main thing is that we understand each other. English is a live language, constantly absorbing new words and constructions, which gives it both depth and power, as well as some kind of universality.

GAN: But this should be done discerningly. It makes perfect sense to include words that bring richness of meaning or invigorate aesthetic expression. It makes no sense to accept elements that dilute potency and flow. While walking to your office, I overheard two women chatting in the hall. "I think he's, like, mocking me, you know?" one of them was saying. "So I'm like, 'Why d'you say that? I mean, 's not cool,' and he's like, 'Oh, I just felt like it,' and I'm like, 'Well, I mean, you should be, like, more polite, you know,' and he's like, 'I *am* being polite," and I'm like, 'Huh?' So I grab my purse, like, and get out of there. I mean, he's not the last man on Earth, like, you know?"

LIN: Did you not understand what she was saying?

GAN: Of course I did. But it brought tears to my eyes. Henry Higgins[5] must have been turning in his grave. The word 'like' was used properly only once. Four times it was used as a filler, along with 'I mean' (three times) and 'you know' (three times), and five times in place of 'I said' and 'he said'. Fillers show poverty of expression, as do the substitutes for specific action descriptors. Students should be encouraged to speak correctly, for sooner than they think, they will face job interview panels where their chances will be greatly enhanced if they show good manners and good language skills. You don't want human oral exchanges to degenerate into strings of empty words and grunts, caveman-style, do you?

LIN: You are too inflexible. Language is what people make of it. It is created by usage. If a word is inserted into free speech often enough by sufficiently many individuals, it becomes part of the language and therefore acceptable. It's a natural process.

---

[4] David Hilbert, German mathematician, 1862–1943.

[5] The phonetics professor in the play *Pygmalion* by George Bernard Shaw (Irish playwright, 1856–1950).

GAN: Should we, then, fully embrace the profane 'f'-word and use it in newspaper articles, television newscasts, and everyday conversation? There is no denying that this word has been adopted by a large part of the population, being heard from some every few seconds. Where do we draw the line if we are to avoid descending all the way into the lexical gutter? I mean, you, Dr. Lin, and your professional brethren should, like, chill out and be a little less liberal in your attitudes toward language molesters, you know? That'd be real cool and, like, awesome, or whatever.

"Lin's logic is deeply flawed," JJ said. "She didn't construct a balanced argument, only one side of it—the side that suited her. A new word, or a new meaning of an older one, becomes formally acceptable when reputable dictionaries include and define it. What could such learned sources write, for example, about the wanton use of 'like' if they were to insert it on their pages?" And JJ recited:

Like. Meaningless word used frequently and at random by the vocabulary-challenged, indicative of verbal negligence and lack of conversational skills.

"For condoning the mutilation of the language," JJ went on, "Lin and others like her are guilty of a kind of illogical behavior that violates the Sixth Mathematical Commandment and brings their MICQ down to 29 or less. In this, I'm afraid, they are on the unBRITE side of things."

"Does Gan have students who make improper use of 'like'?"

"One or two. He tries to show them the error of their ways by resorting to an analogy. How would they feel, he asks, if mathematicians developed such a habit in *their* lingo? What if they took to saying 'pi' whenever and wherever they pleased while lecturing? Here is a 'clean' spoken sentence and its versions contaminated by 'like' and then by 'pi':

The fraction sine of x over x tends to one as x tends to zero.
The fraction sine of, like, x over x tends to one as x tends to, like, zero.
The fraction sine of, pi, x over x tends to one as x tends to, pi, zero.

Apart from sounding horrible, Gan says, the last one can also be taken to imply that $\lim_{x \to \pi 0} (\sin(\pi x))/x = 1$, which is wrong. Mathematics is a science of precision and careless language has no place in it."

"What kind of reaction does he get from the students?"

"They tell Gan that they want to kick the habit but don't know how. So he offers them a simple solution, based on the example of Demosthenes[6] coupled with what Pavlov[7] called a conditional reflex. He recommends that when they come down to breakfast every morning, they should repeat 'like' one hundred times, loudly and uninterruptedly, in front of their family."

---

[6] Greek orator, 384–322 BC, who, according to legend, cured himself of a speech impediment by placing stones in his mouth and talking over the roar of the sea waves.

[7] Ivan Petrovich Pavlov, Russian physiologist and physician, 1849–1936.

"And that helps?"

"They would embarrass themselves and annoy everybody around. At the same time, their brain would get saturated with the sound of the word, which, when heard again later, would immediately be recognized and associated with unpleasant feelings. This should make it easier for them to start reducing the frequency of 'like' in their own speech and ultimately eliminate it altogether."

"Hmm.... One hundred times, you say. Why exactly one hundred?"

"Because ninety-nine is too short, whereas a hundred and one would be a bit over the top."

I looked at JJ to see if his weird sense of humor was accompanied by the sketch of a smile: not a chance. All I saw was his large dark eyes and a face as stern as ever.

"You'll be talking about grammatical errors next," I said.

"Many don't have a clue what grammar is and don't care. How often, when you greet people with 'Hello! How are you?' they reply, 'Hi! I'm good'?"

"Every day."

"This would be a legitimate answer if they were, say, yummy candy bars or saintly persons. But you aren't asking them how they taste or if they are of good or bad character. You want to know how they *feel*, so they should say 'I'm well', or 'Fine', or 'I'm a little under the weather', or 'What's it to you?', or something similar. Adjectives and adverbs have different roles in a sentence. 'Good' is an adjective. If we are to use it as an adverb, then we should also be allowed to say 'I want my steak good done' or 'We need to make a good-chosen substitution', but I'm sure people would find such formats unacceptable. Talking about which, Gan once tried in vain to explain a point of syntax to Stu. Here is their VOTSITed conversation."

STU: Why did you underline this sentence in red on my test, Dr. Gan? It's very clear what it means.

GAN: Its meaning is clear, but its construction is wrong.

STU (puzzled): It just says, "Replacing formula (1) in equality (2), the equation takes the form (3)." I see nothing wrong with it.

GAN: You've got there what is known as a *dangling participle*.[8] The clause 'Replacing formula (1) in equality (2)' is without a subject. Who is doing the replacing?

STU: Me, naturally.

GAN: You should use 'I', not 'me', in this instance. Well, if *you* are doing it, then why not say so? You could choose any one of a number of equally valid alternatives. The simplest would be to write something like "We[9] replace formula (1) in

---

[8] Or 'hanging gerund', in British English.

[9] This is not the royal 'We'. It is the style adopted by scientists when they write learned articles and books. The explanation for its usage is unclear. It could be that 'I' might sound too boastful and arrogant; or that 'we' bestows more prestige on them; or that they want to make the reader feel included; or none of the above.

equality (2) and bring the equation to the form (3)." Or, less verbosely, you could say "The replacement of (1) in (2) brings the equation to the form (3)."

STU (digests all this): Dr. Gan, is my solution to the problem correct?

GAN: Yes, it is.

STU: Then how will this dangling participle make me a bad engineer? How would it prevent me from building a sturdy bridge or a fine-tuned car engine? Between you and I, it doesn't matter whether my clause has a subject or not.

GAN: You should use 'me', not 'I', this time, because it is governed by a preposition. I'm sure you'll be a fine engineer, Stu, but you won't be a true intellectual unless you learn to present your correct solutions in an equally correct spoken or written form. What you jotted down in your test is like serving chocolate cake out of a newspaper. Or like wearing a sweater back to front: it still does its job by keeping you warm, but it says something not very complimentary about you.

"Gan has heard students use 'maxima', 'minima', 'extrema', 'data', 'bacteria', 'phenomena', and 'criteria' as words in the singular," JJ added. "He always points out to them that these are plurals and that their corresponding singular forms are, respectively, 'maximum', 'minimum', 'extremum', 'datum', 'bacterium', 'phenomenon', and 'criterion', with the first five coming from Latin and the last two from Greek."

"While we're on this subject, I'm sure you know the plural of 'lemma'?" I asked.

"A good and subtle one! It's 'lemmata'. Another Greek word. However, almost everybody uses 'lemmas' instead." JJ seemed to take a deep breath. "Regarding mathematical terms," he continued, "since language is an essential ingredient in the buildup of mathematical ideas, do you think that mathematical expression should, in turn, be looked upon as some kind of linguistic manifestation?"

"What do you mean?"

"Gan once discussed the similarity between mathematics and language with one of his female students, Fem, and her male colleague, Mal. Have a look," JJ said, and activated his well-concealed VOTSIT.

GAN: Experts claim that language has algebraic features. Have you noticed any?

FEM: Well... let's consider the sentences *Tom chased the cat* and *The cat was chased by Tom*. They bring to mind an algebraic equation of the form $x = a + y$, in which x refers to Tom, y to the cat, and a to the verb 'chase'. If you solve it for y, you get the equivalent form $y = -a + x$, which corresponds to the second sentence and in which $-a$ represents the passive form of the verb."

MAL: An interesting pattern-sharing. You think logic can also be used to analyze other structures, Dr. Gan?

GAN: Sometimes. For example, the correct positioning of words in a sentence. Take the word 'only': it can be inserted in the sentence *Mary drinks water at lunch* in four different places, each time generating a different meaning:

Only Mary drinks water at lunch. (Nobody else does that.)

Mary only drinks water at lunch. (All she does to water is drink it.)
Mary drinks only water at lunch. (She drinks nothing else.)
Mary drinks water only at lunch. (That's the time when she does it.)

Unfortunately, some people use the second form to convey the meaning of the fourth one.

FEM: But that's incorrect. It's like inserting, say, the figure 4 in the number 1762 between 1 and 7 when you really want it between 6 and 2. It's silly to say 14762 and expect people to understand 17642.

GAN: Precisely.

MAL: Should we, perhaps, look at the analogy differently? Not as 14762 versus 17642, but as $1 + 4 + 7 + 6 + 2$ versus $1 + 7 + 6 + 4 + 2$, which yield the same result?

GAN: Nice try. It wouldn't work, though, because this would lead to the conclusion that 'only' could be inserted anywhere and the meaning of the sentence would remain the same in all cases.

FEM: It seems that this type of analogy has severe limitations.

GAN: You're right. Here we don't have an 'if and only if' equivalence, but rather a unidirectional implication, so to speak. As Truesdell[10] wryly observed, "There is nothing that can be said by mathematical symbols and relations which cannot also be said by words. The converse, however, is false. Much that can be and is said by words cannot successfully be put into equations, because it is nonsense."

"Anything on spelling?" I asked.

JJ waved a hand. "That would take forever. But I'll mention punctuation, which, like grammar, is a matter of logic. Don't you just love it when you read in a test answer that 'the function is differentiable, *it's* derivative exists and *its* calculated by means of known rules'? Why is it so hard to remember that *its* is the possessive pronoun whereas *it's* is the contracted form of 'it is'?"

"At least such errors don't show in speech. Same about the misuse of the Saxon genitive."[11]

"Don't go there," JJ rumbled ominously. "I'm sick and tired of seeing people write 'Maxwells[12] equations' instead of 'Maxwell's equations'. Last winter, on a university campus—a *university* campus, a place of higher learning—I noticed a prominently displayed sign bearing the inscription 'Seasons greetings! May it's blessings be with you!' How could all those academics walk by and not shudder? If they accept such errors, what hope is there for the likes of Bob Jr.?"

"Who's Bob Jr.?"

"The guy who owns the fast-food chain called Bob's Jr. He doesn't know that 'Bob's Jr.' means 'the junior of Bob', and that to indicate the joint owned by Bob Jr. he should've written Bob Jr.'s."

---

[10] Clifford Ambrose Truesdell III, American mathematician and natural philosopher, 1919–2000.

[11] Saxon genitive is the term that describes the English possessive '*s*.

[12] James Clerk Maxwell, Scottish mathematician and physicist, 1831–1879.

"Punctuation is just a convention, if you think about it."

"But any good convention should be justifiable. Here is an example that irritates me badly. It's to do with the positioning of punctuation signs in relation to quotation marks," JJ said, and hurriedly scribbled on a white paper napkin:

(A) John told Mary to 'have a word with the top man', within a day or two, because the issue was 'urgent and delicate'.

(B) John told Mary to 'have a word with the top man,' within a day or two, because the issue was 'urgent and delicate.'

(C) John told Mary, 'You should have a word with the top man, within a day or two, because the issue is urgent and delicate.'

"You think these sentences are punctuated correctly?"

I shrugged. "Yes, all three of them. (A) is correct in British English, (B) in American English, and (C) in both."

JJ seemed disappointed that he hadn't caught me out. "You must admit, though, that only (A) and (C) are backed by logic. (C) is simple to explain: the quote is a full sentence and the final period is part of it, so it goes with the sentence inside the quotation marks. Reproducing only a brief section of someone's speech is an entirely different matter. Those words are placed in inverted commas and treated as a solid block. Any punctuation sign that ends or breaks the sentence after the block should be inserted outside the closing quotation mark because the entire block, including the last quotation mark, is part of the sentence or clause being punctuated. The American convention used in (B) is not only illogical, but also discriminatory in the sense that it applies exclusively to the period and the comma. Why not extend it to all the punctuation signs that fragment the narrative? If constructions like

He called it 'strange.'
He called it 'strange,' I think.

are mandatory in the U.S., then so should be the following ones:

He called it 'strange?'
He called it 'strange!'
He called it 'strange:' it was his idea.
He called it 'strange;' perhaps it was.
He called it 'strange...'
He called it 'strange—' was it?—and it stuck.
He called it something (probably 'strange)' and it stuck.

Total nonsense, of course! In the same vein, how would you feel if this silly convention were carried over to mathematical texts? If, for example, operation blocks delineated by parenthesis were breached by punctuation signs? I bet that a printed sentence such as

The solution is given by the simple sum $2 + (5 - 3.)$

would make your blood boil."

"It certainly would."

"What's done in sentence (B), twice, is like stuffing a shirt in a drawer with one of its sleeves hanging out. I have no idea who came up with this harebrained style, but I'm surprised to see it still in use. It's sloppy. It's unsightly."

"It's indefensible," I added, for good measure.

"Lack of logic of the first magnitude." JJ leaned back in his chair. "Now," he said, "it's time for your stories."

I finished the last of the coffee and pushed the empty cup aside. "OK, here we go. Today I've got three. First one coming up."

Miss Jones looks at her students and says, "Tracy, go to the map and find North America."

The girl walks to the far wall and points, "Here it is, Miss."

"Very well. Now, everybody: who discovered America?"

Throwing their hands up, all students shout enthusiastically, "Tracy did, Miss!"

"Enough of that!" Miss Jones snaps at the class. "We'd better see if you learned the words that I gave you yesterday. Paddy, how do you spell 'crocodile'?"

"C-r-o-k-a-d-i-a-l."

Miss Jones sighs. "Young man, that's completely wrong."

"Perhaps, Miss. But that's how *I* spell it."

The teacher is about to reply when she spots a boy unwrapping a candy bar and biting off a big chunk of it. "Bubba," she remonstrates, "I think this is the fifth one in a row! It's not good for your health. Your teeth will rot and you'll get fat. Besides, you are not allowed to eat or drink in the classroom."

"My grandpa lived to be 101," Bubba says, munching away.

Miss Jones feels irritated. "And that's because he ate a lot of candy bars?"

"Nah!" Bubba snorts. "That's because he minded his own business."

"Just as Gan recorded earlier," JJ noted. "The shallowness of public school education and the students' lack of respect for their instructors."

I let that go. "Here's the second one, which also mentions spelling."

A woman walks into an ice cream shop and tells the attendant, "I'd like a large chocolate cone, please."

"I'm sorry, ma'am," the attendant replies, "we are out of chocolate."

"In that case," the woman says, "please let me have a small chocolate cone."

The attendant shakes his head. "As I explained, we have no more chocolate today. We've run out of it. But I could give you another flavor."

"That's too bad. Well, then give me a small cup of chocolate ice cream."

Exasperated, the attendant says, "We don't seem to be getting very far with this conversation, do we? So let's try something different. Can you spell 'van', as in 'vanilla'?"

"Sure: v-a-n."

"Great. Now can you spell 'straw', as in 'strawberry'?"

"It's s-t-r-a-w."

"Excellent! Then how about spelling 'stink', as in 'chocolate'?"

The woman looks baffled. "But there is no 'stink' in 'chocolate'."

"Exactly!" beams the attendant. "That's what I've been trying to tell you for the past two minutes: there is no stinkin' chocolate!"

"The third one," I said before JJ had time to comment, "is of an entirely different nature. Since you mention your surveys so often, here's a story involving one of our own."

A pollster is sent out to determine how much scientists and engineers really know about mathematics. A single question is asked: is it true or false that all odd integers greater than 1 are prime numbers?[13] The answers received vary with the specialism of the interviewee.

Chemist: "What is a prime number?"

Physicist: "3 is prime, 5 is prime, 7 is prime, 9 is—hmm...well, it's not prime, but you always have experimental errors—11 is prime, 13 is prime, so all odd integers greater than 1 are prime."

Psychologist: "3 is prime, 5 is prime, 7 is prime, 9 is prime but is trying to suppress it, 11 is prime, 13 is prime, so all odd integers greater than 1 are prime."

Political scientist: "I'm told that some of them are prime, but we should aim to create a more equitable society where all numbers have a fair chance to develop their potential and become prime."

Engineer: "3 is prime, 5 is prime, 7 is prime, 9 is prime, 11 is prime, 13 is prime, 15 is prime, 17 is prime, 19 is prime, so all odd integers greater than 1 are prime."

"Many of these scientists bad-mouth mathematicians themselves, so I suppose it's fair." JJ put his hands on the table. "Thanks for the stories," he said. "I've really enjoyed our little chat. If all goes well, we'll talk some more next time." Then, unceremoniously, he sprang to his feet and promptly walked out of the bar.

Notes After the Meeting

Here are the answers to JJ's questions together with some brief personal comments.

Solution to Question 1. This is a very easy problem if we look beyond its elaborate and dramatic setup (and accept that there are flies capable of such exploits). To be precise, we are dealing with an object (the fly) moving in a straight line at a constant speed. As we know, in a motion of this type,

$$distance = speed \times time.$$

The speed of the object is 200 km/h. The duration of the object's motion is the time taken by the two locomotives to crash into each other. The locomotives

---

[13] A prime number is a positive integer that divides exactly only by 1 and itself.

are approaching at a relative speed of $150 + 100 = 250$ km/h. The crash occurs when, between them, they have covered the total distance of 500 km that separated them originally, that is, after $500 \div 250 = 2$ h. Since the whole process, start to collision, takes 2 hours, the fly has traveled a total distance of $200 \times 2 = 400$ km.

"Why do you have to kill a fly to prove a mathematical point?" somebody once asked me. "Why are you so cruel to harmless little beasties? Can't you use an abstraction instead, something like a 'flying particle'?"

After due consideration, I said that an abstraction would sound incongruous in a problem of this nature, that a fly with a perfectly rectilinear trajectory was just the thing to be squashed between two crashing locomotives, and that, for purely personal reasons, I was a lifelong avowed enemy of those harmless little beasties and supported all commercial outlets that stored fly-exterminating devices on their shelves.

On a more serious note, this is another one of those problems where the simplest solution is obscured by a contorted narrative. If you did not think of this solution and believed that a very long chain of computations would be necessary to arrive at the final answer, don't be embarrassed: you are in good company, for this also happens from time to time to seasoned academics, as the next anecdote shows.

It is said (apocryphally, perhaps) that John von Neumann[14] was a guest at a private party when the host, to liven up the atmosphere, asked everybody to try to solve the problem with the fly and the locomotives.

No sooner had the host finished explaining the problem than von Neumann gave the correct answer. As you can imagine, this annoyed the host. "Oh, what a shame," he said reproachfully. "You knew the problem already."

"I can assure you that I didn't."

"Then you must've seen the short solution very quickly," the host insisted. "Either way, I wish you'd kept quiet."

"What short solution?" the other guests asked.

"Well, you see," the host said, "on hearing the problem, people normally follow their intuition and rush to compute the distance traveled by the fly in discrete individual steps. Thus, they find (i) the position of the engine coming from Y when the fly meets it for the first time, then (ii) the position of the engine coming from X when the fly, going back toward X, touches its front end, and so on, constructing an infinite geometric series that sums up to the distance required. But they don't need to go to all that trouble." And the host explained the simple method connecting the total flying time, speed, and distance. "I'm really disappointed in you, sir," he added, turning to von Neumann again, "for giving the game away so quickly."

John von Neumann looked sheepishly into his glass of wine. "I'm sorry," he mumbled. "What you call the short method never crossed my mind. Myself, I just

---

[14] Hungarian-born American applied mathematician and one of the creators of computer science, 1903–1957.

summed up the infinite series."

If you want to know how to do that long step-by-step computation, here is the answer in terms of more general parameters.

Suppose that the top speed of the engine at X is u km/h, that of the engine at Y is v km/h, and that of the fly is w km/h, where $w > u$ and $w > v$, and let $XY = d$ km. We write

$$\alpha = \frac{(w-u)(w-v)}{(w+u)(w+v)}, \quad \beta = \frac{2dw^2}{(w+u)(w+v)}. \qquad (5.1)$$

If you perform the gory details, you will find that the first leg traveled by the fly (toward Y) is β, the second leg (back toward X) is $\beta\alpha$, the third one (again toward Y) is $\beta\alpha^2$, and so on; therefore, the total distance is

$$D = \beta + \beta\alpha + \beta\alpha^2 + \cdots = \beta(1 + \alpha + \alpha^2 + \cdots).$$

The expression in brackets is an infinite geometric progression with common ratio α. Since, as seen from (5.1), $0 < \alpha < 1$, this series is convergent and the right-hand side above is equal to

$$\beta \frac{1}{1-\alpha} = \frac{2dw^2}{(w+u)(w+v)} \times \frac{1}{1 - \dfrac{(w-u)(w-v)}{(w+u)(w+v)}} = \frac{dw}{u+v};$$

therefore,

$$D = \frac{dw}{u+v}. \qquad (5.2)$$

In our case, formula (5.2) with $u = 150$, $v = 100$, $w = 200$, and $d = 500$ yields $D = 400$, as expected.

Occasionally, a few students give the correct answer by the short method, but I have a sneaky suspicion that, unlike John von Neumann, they had heard the problem before.

Solution to Question 2. Consider the number

$$n = \sqrt{2}^{\sqrt{2}}.$$

This is a real number, so it is either rational, or irrational.

If n is rational, then we have an irrational power ($\sqrt{2}$) of an irrational number ($\sqrt{2}$) that is rational.

If n is irrational, then

$$m = n^{\sqrt{2}} = \left(\sqrt{2}^{\sqrt{2}}\right)^{\sqrt{2}} = \sqrt{2}^{(\sqrt{2})^2} = \sqrt{2}^2 = 2,$$

which is rational; therefore, we again have an irrational power ($\sqrt{2}$) of an irrational number (n) that is rational. Hence, it seems that the the original question can be answered in the affirmative.

Although apparently flawless, this solution has something in it that makes one feel uneasy. When playing with the number n, we reasoned smartly in 'ifs' and 'thens' but did not discover whether n is rational or irrational. We just speculated about its nature in a convenient and logical way. Because of this, certain specialists in set theory would argue that the correct formulation of the problem should include a supplementary clause: "Show that it is possible for some irrational power of some irrational number to be rational, *provided that the question is decidable.*"

Mathematics is seldom as straightforward as it looks.

A Word to the Wise

The Number System

In the beginning, there were the natural numbers (positive integers):

$$\mathbb{N} = \{1, 2, 3, \ldots\}.$$

These are the numbers used for counting; they can be added and multiplied without restriction, which is a very good thing. Subtraction and division, however, can be performed only with some pairs of such numbers.

The set $\mathbb{N}$ was then augmented with 0 and the negative integers to become the set of integers:

$$\mathbb{Z} = \{\ldots, -3, -2, -1, 0, 1, 2, 3, \ldots\}.$$

The operation of subtraction is now possible for all pairs of numbers in the set.

Later, the larger set $\mathbb{Q}$ of all rational numbers was introduced—the set of all numbers that can be written in the form p/q, where p and q are integers, $q \neq 0$, and the highest common factor of p and q is 1.

The set $\mathbb{Q}$ permits all four fundamental arithmetic operations. Division by 0, however, is not allowed.

Finally, there came the construction of the even larger set $\mathbb{R}$ of real numbers. These are all the numbers that can be put in a one-to-one correspondence with the set of points on a line. A real number is either rational or irrational. Examples of irrational numbers are $\sqrt{2}$, π, e, etc.

Aside from possessing the freedom of all arithmetic operations, this set also allows limiting processes to become meaningful. Alas, division by 0 remains banished to this day.

The set $\mathbb{R}$ includes every single number that makes sense in the physical universe. But since, when it comes to science, mathematicians are a greedy lot, they went on to invent other types of numbers, such as complex numbers, which, although artificial constructs, prove to be invaluable tools in the solution of many practical problems.

SCAM 6

Foreign Languages

Rident stolidi verba latina.[1]

Publius Ovidius Naso[2]

Wer fremde Sprachen nicht kennt, weiß nichts von seiner eigenen.[3]

Johann Wolfgang von Goethe[4]

I met JJ again at a mathematics conference sponsored by Florida State University in Tallahassee, Florida. Just like the first time, he was sitting alone at a table in a corner of the hotel bar, gazing at me across the room with dark, inscrutable eyes.

As soon as I sat down, he pushed a piece of paper in my direction. "Here," he said laconically. "For your students."

I glanced at the paper and smiled. The text read something like this.

Question 1 (requires complex numbers). *What, if anything, is wrong with the following assertion and proof?*

Theorem. *Global warming is way out of hand.*

Proof. The demonstration consists of two parts.

(i) First, we show that any negative number coincides with its positive counterpart. Let $-c$ be any negative number (which means that $c > 0$) and consider the obvious equality

$$\sqrt{-c^2} = \sqrt{-c^2},$$

which we can also write as

$$\sqrt{\frac{-c^2}{1}} = \sqrt{\frac{c^2}{-1}},$$

or

$$\frac{\sqrt{-c^2}}{\sqrt{1}} = \frac{\sqrt{c^2}}{\sqrt{-1}}. \tag{6.1}$$

---

[1] Fools laugh at Latin. (Latin)
[2] Roman poet, 43 BC–AD 17; known as Ovid.
[3] He who does not know foreign languages, knows nothing about his own. (German)
[4] German poet, novelist, playwright, and natural philosopher, 1749–1832.

In terms of complex numbers (see p. 65), (6.1) is the same as

$$\frac{ci}{1} = \frac{c}{i}, \qquad (6.2)$$

where $i^2 = -1$. Cross-multiplying in (6.2) now leads to

$$ci^2 = c,$$

and we deduce that $-c = c$, as stated.

(ii) We now go back to the assertion of the theorem. According to part (i), any subzero temperature recorded on Earth is equal, in fact, to its absolute value, which means that there are no winters to speak of anywhere on the planet and that the polar regions themselves are enjoying a hot tropical climate. This is a catastrophic global superwarming phenomenon that so far seems to have escaped the notice of the scientists.

Question 2 (requires basic arithmetic or algebra). *John and Mary are children in the same family. John has as many brothers as sisters. Mary has twice as many brothers as sisters. How many boys and how many girls are there in that family?*

"Interesting," I said. "Thanks." I folded the paper and put it in my pocket, waiting for JJ to broach, as usual, a subject of his choice. He needed no special invitation.

"The Anglo-Saxon world," he began abruptly, "seems to have scant consideration for other languages. Not only do some of its inhabitants expect everyone else to speak English when they travel abroad, but in rare cases of extreme arrogance they even demand it. The primary reason for this has to be their incomplete education. Modern languages are not cool, like the history of rock-and-roll or video game techniques, so very few bother to study how other people speak and think."

"Can you blame them? Almost anybody in the world manages some English these days. It is the planet's *lingua franca*."[5]

"Careful," JJ noted wryly. "You've just used another language."

"And people tend to be utilitarian. When something works, they are reluctant to interfere with it."

"In business, using a single 'universal' language is very advantageous. But not being able to speak or understand other tongues, however little, can be culturally harmful. In my view, one's stern refusal to accept this fact is illogical. Since individuals in that position contravene the Sixth Mathematical Commandment—thou shalt always be guided by logic in thy thinking and actions—they are unBRITEs with a MICQ of at most 29."

---

[5] In Italian, *lingua franca* literally means Frankish tongue. (The Arabs, who used the term originally, called all the inhabitants of Western Europe 'Franks'.) In a wider sense, it refers to a language adopted as a medium of communication between people speaking different languages.

"Has your friend Gan been in any unusual situations involving languages?" I baited him.

"He has. Watch this classroom episode, recorded on his VOTSIT, where he converses with Stu."

GAN: Can you tell me why the boundary value problem on the board has only the zero solution?

STU (confidently): Because we have the Laplace[6] equation with homogeneous Dirichlet[7] boundary conditions.

GAN: Correct mathematical reason; wrong name pronunciation.

STU (nonplussed): Dir-*ish*-lay? What's wrong with that?

GAN: It's Dir-ih-*çleh*.[8]

STU: But wasn't he French?

GAN: Some people tend to assume so. Dirichlet was born in Düren, a German town that at the time belonged to the French empire, and lived the greatest part of his life in Germany. His grandfather had come from the Francophone area of Belgium, which explains 'Lejeune' among his names and, in 'Dirichlet', the mute final 't' and the accent on the last syllable. At the same time, the 'ch' is pronounced as in the German word 'dich', not as the French 'sh'. Unfortunately, a lot of American, British, and Australian mathematicians insist on perverting these elements, saying 'Dir-*ish*-lay', which is wrong in both sound and accent, while the rest of the world, including the French and the Germans, use the correct pronunciation of 'Dir-ih-*çleh*'. To avoid confusion, you should know that, as a rule, a person's name is best pronounced the way the person himself or herself pronounces it.

"There are many cases like this," JJ continued. "For example, I've also heard 'Lagrange'[9] pronounced 'Lagrandjee' instead of 'La-*granzh*', 'Rolle'[10] pronounced 'Rollee' instead of 'Rohl', 'Schwarz'[11] pronounced 'Shuartz' (as in 'quartz') instead of 'Shvartz', 'Minkowski'[12] pronounced 'Mincowski' instead of 'Min-*cov*-ski'—the list goes on and on. When asked why they do it wrongly, the perpetrators often say that it doesn't really matter because those characters are already dead. I don't think that's a very good answer, do you?"

---

[6] Pierre-Simon, Marquis de Laplace: French mathematician and astronomer, 1749–1827.

[7] See footnote on p. 28.

[8] The symbol 'ç' in the German phonetic alphabet denotes a sound similar to that of a drawn-out 'h' in 'huge'. A much coarser approximation, but perhaps easier to reproduce, would be the 'ch' in the Scottish 'loch'.

[9] Joseph-Louis Lagrange: Italian mathematician, baptized as Giuseppe Lodovico Lagrangia, who lived the later part of his life in France, 1736–1813.

[10] Michel Rolle: French mathematician, 1652–1719.

[11] Karl Hermann Amandus Schwarz: German mathematician, 1843–1921.

[12] Hermann Minkowski: German mathematician, 1864–1909.

"Sometimes we have great difficulty with foreign names, so we do the next best thing: we 'approximate' them in English."

"That's less damnable. What really saddens me is to see people born in other lands who themselves allow their names to be anglicized because the English speakers around them are too inexpert at languages or don't want to learn their proper pronunciation. Austrians called Loewe who do not bristle up when they hear 'Low' (correct: $Lœ$-veh) or Poles called Nowacki who turn a deaf ear to 'No-wacky' (correct: Noh-*vats*-kih) should show some backbone and make a stand for it. They ought to emulate Professor Ath from Athens: incensed by the ill-treatment of the Greek letters by foreign mathematicians, he has compiled a table of the most used ones which he distributes freely at international conferences. I've got a copy for you."

And JJ pushed another piece of paper in my direction. The text on it was neatly laid out, as shown below.

Symbol	Name	Greeks say	Others (wrongly) say
α	alpha	*ahl*-fah	*al*-fah (as in 'album')
β	beta	*beh*-tah	*bay*-tah, *bee*-tah
γ, Γ	gamma	*gahm*-mah	*gam*-mah (as in 'gamble')
δ, Δ	delta	*del*-tah	triangle
ε, ϵ	epsilon	*ep*-see-lohn	ep-*sy*-lon (as in 'nylon')
ζ	zeta	*zeh*-tah	*zay*-tah, *zee*-tah
η	eta	*eh*-tah	*ay*-tah, *ee*-tah
θ, Θ	theta	*theh*-tah	*thay*-tah, *thee*-tah
ι	iota	*ioh*-tah (as in 'York')	eye-*o*-tah
$\kappa, \varkappa$	kappa	*kahp*-pah	*cap*-pah
λ, Λ	lambda	*lahmb*-dah	*lam*-dah, *lahn*-dah
μ	mu	mu	
ν	nu	nu	
ξ, Ξ	xi	xee	zye
π, Π	pi	pea	pie
ρ	rho	roh	roe
σ, Σ	sigma	*sig*-mah	
τ	tau	tau (as in 'how')	taw (as in 'raw')
υ, Υ	upsilon	*ewe*-psih-lohn	*up*-silon
ϕ, φ, Φ	phi	fee	fye
χ	chi	khee	kye
ψ, Ψ	psi	psee	sye
ω, Ω	omega	oh-*meh*-gah	o-*may*-gah, *ohm*-eh-gah

"Ath," JJ went on, "makes one concession: π (and its upper case version Π) can be mispronounced 'pie', to avoid the vulgar connotation of the original sound. He also points out that the pronunciation of the letters as symbols in an academic context, listed in the third column of his table, is that of classical, not modern, Greek."

"How do the British fare in this?"

"They are elated to see English replace French as the dominant international language in business, culture, and diplomacy. They view this as just retribution for the fact that the inscriptions on the British Royal Coat of Arms are in French: *Honi soit qui mal y pense*[13] and *Dieu et mon droit*.[14] Otherwise they are relaxed about the whole thing. Having emerged from their empire period, during which they taught English to a large part of Earth's population, the British are now undergoing a positive change of attitude toward things foreign. They travel extensively on holiday to mainland Europe and, as a result, many have learned to say quite cleanly *Una cerveza, por favor*,[15] or *Ein Wurst mit Brot und Senf, bitte*,[16] or *La nostra squadra di calcio è migliore della vostra*.[17] Why can't you do the same?"

"Europe is a lot further away for us. We have less opportunity to flex our feeble linguistic muscles."

JJ would not relent. "Yet American English keeps adopting words and expressions from other languages. But more often than not it gets the wrong end of the stick and skews their original meaning or pronunciation. Have a look at some examples gathered by Gan from exchanges with various serving personnel, Ser, in restaurants."

GAN (helps himself to a piece of buffet steak): Nice beef you've got here.

SER (points to a pot of gravy beside which there is a label with the inscription 'Au jus'): Thank you, sir. Would you like some oh-juice with that?

GAN (benevolently): Ma'am, you've made a double error. First, *au jus* is a French expression that literally means 'with gravy'; the word 'with' in your question is therefore redundant. And second, *jus* is pronounced 'zhu' in French, not 'juice'.

SER (shrugs with indifference): Sir, in my joint this is an American expression and it's 'oh-juice'. Do you want it on your steak or not?

GAN (taking a principled stand): I assume that it's been prepared with the same skill you have in foreign languages, so I'll pass.

SER (notebook and pen in hand): Have you decided which entrée you would like?

GAN: Why do you refer to a main dish as an 'entrée'? This is a French word that means 'entrance' or 'input'. In European culinary parlance it designates a dish served at the beginning of a meal.

SER: So what should I call it?

GAN: Main dish, or main course.

---

[13] *Shame on him who thinks this evil.* The motto of the ancient Order of the Garter.

[14] *God and my right.* The motto of the British Sovereign, dating from the 15th century.

[15] A beer, please. (Spanish)

[16] A sausage with bread and mustard, please. (German)

[17] Our soccer team is better than yours. (Italian)

SER: Can't do that. If my boss hears me, I'll get fired. He told us that the first rule to remember while working in his restaurant is not to be a smart aleck. Since I need this job to pay my college fees, I've got to go by what he says. (Lowers his voice.) *Mais je peux chuchoter 'plat principal' pour vous.*[18]

GAN (elated): Here is a twenty-dollar bill for your tip, young man. I'm glad to see that you are one of us, even if you do work undercover.

SER: Welcome to our new restaurant, Nouveau American Cuisine.

GAN: Do you realize that this name is wrong?

SER (recoils a little): Wrong?? We've advertised it extensively on the TV and got no complaints so far.

GAN: You obviously chose the name because it sounded French and thought that anything French about food could only be great and enticing. You didn't bother to check if the expression was legitimate. In fact, the expression is a horrible hybrid of words put together with a pitchfork by a semidoct in a hurry. In French, words have gender, and since *cuisine* is feminine, the proper text should have read 'Nouvelle cuisine américaine'.[19] Or, if you insist on mongrelizing it, 'Nouvelle American Cuisine'.

SER: Here is your calzone, sir.

GAN (patiently): It isn't 'cal-*zone*'; it's cal-*tzoh*-neh'. Why don't you pronounce it like the Italians? After all, they invented it.

SER: Because this is how it is spelled on the pack: cal-*zone*.

GAN: Do you think that mathematics is the most important science in the world?

SER (dismissively): Nah. That's astrology. Why?

GAN: Just checking. As I suspected, your MICQ is at most 40. Although—astrology?—that's far too optimistic.

"The British," JJ said, "are also up on you in the correct pronunciation of words like 'Italian', 'Iran', and 'Iraq'. In America one often hears 'Eye-talian', 'Eye-ran', and 'Eye-raq'. If you thought this came only from JOE in the street, you'd be mistaken. Pundits on the radio and TV, who should know better, are also guilty of this cultural misdemeanor. But wait a minute: I'm talking about the media, so I should not mention culture in the same sentence. I dislike oxymorons."

"Anyone else you feel like bashing today?"

"Yes. Some backwoods lawyers. In English law the French expression *voir dire*, pronounced 'vwahr deehr', describes an investigation during a trial which intends to establish if the evidence given by a witness is true or admissible.[20] In the U.S., whose legal system is based on the English one, it means probing the

---

[18] But I can whisper 'main course' for you. (French)

[19] The French don't capitalize every word in a title, or the adjectives formed from countries' names.

[20] The Romans had the equivalent expression *veritatem dicere*, which, like the French one, translates as 'to tell the truth'.

suitability of a juror. Not quite the same thing, is it? What truly intelligent person would knowingly corrupt the meaning of a legal term or accept such a vitiated term as part of their vocabulary? To make matters worse, in some states the phrase is pronounced 'vor-*dire*'. Clearly, not all lawyers are smart, well-educated professionals. But why should I be surprised that people have so little knowledge of foreign languages when they are misusing their own?"

"How's that?"

"You call *football* a game where the ball is controlled, passed, and carried almost exclusively by hand, being kicked only once in a long while, and where a scoring play is called a *touchdown* although the player involved is not compelled to touch anything down, up, or sideways."

"Does Gan discuss languages in the classroom?" I asked, trying to move the conversation in a slightly different direction.

"Frequently. Here is the recording of a recent dialogue with Stu on this subject."

GAN: Many mathematicians seem interested in foreign languages and appear to have less difficulty learning them than other people. What do you think is the explanation?

STU: Could it be their logical mind? Their systematic way of dealing with rules and structures?

GAN: It could indeed. Mathematicians also have the advantage in this respect over the general public since, in the process of learning their scientific jargon, they gain a certain amount of experience with the two main parts of a new language—grammar and vocabulary—and their syntax connector.

STU: Which of those two parts is more important?

GAN: Both are essential. You may know an awful lot of words, but if you don't know the rules for piecing them together meaningfully, you are stumped; on the other hand, you may master all the rules, but if you don't have an adequate vocabulary to apply them to, you won't be able to communicate.

STU: What would you say was the minimum number of words needed to get by in a new language?

GAN: It depends. For a basic, limited chitchat, I guess around five hundred. Considerably more if you want to lecture on the teachings of Socrates.[21]

STU: Does this apply to mathematics as well?

GAN: One can draw surprising analogies between a foreign language and mathematics, where numbers and letters may be regarded as 'words', operational signs and symbols as 'grammatical' markers, and equalities and other relations as 'sentences'. These elements make for great efficiency in expressing mathematical thought. To answer your question, yes, I would say so, although here the minimal figure would depend heavily on the definition of 'chitchat'.

STU: Can you give us a small-scale example?

GAN: How many different words do you normally need in a spoken language to read the numbers on a clock face? Ten, if you choose to say 'one-zero', 'one-one',

---

[21] Classical Greek philosopher, c469–399 BC.

and 'one-two' instead of 10, 11, and 12. To write the same in mathematics you need only one digit—that is, one 'word'.

STU: I've seen such a clock; it uses the digit 9 exactly three times to express each of the clock-face numerals.

GAN: Can you do the same with, say, the digit 1?

STU (a few minutes later, emerging from an animated huddle with his classmates): We've cracked it, Dr. Gan. (Walks up to the board and sketches.)

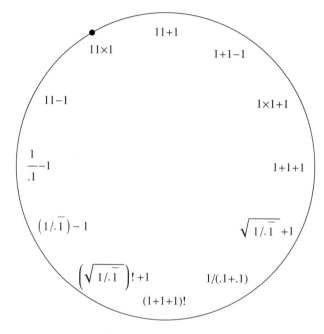

GAN: Excellent, Stu. You made clever use of the factorial in writing 6 as $3!$, and of the well-know notation $.\overline{1}$ for the nonterminating decimal $1/9 = .111\ldots$. Now comes a trickier task: prove that the job can be done with any other nonzero digit n, used, as before, exactly three times to produce each of the numbers from 1 to 12. Remember that literal-symbol operators such as exp, ln, and so on are not allowed; nor are symbols invented by yourselves.

STU (after a longer, even more animated huddle): We give up. We've been able to sort out only
$$1 = n^{n-n}, \quad 2 = (n+n)/n.$$
We don't see how to generate the rest.

GAN: Before I show you a solution, let us recall that for any real number x, the *floor* of x, denoted by $\lfloor x \rfloor$, is the largest integer less than, or equal to, x, and the *ceiling* of x, denoted by $\lceil x \rceil$, is the smallest integer greater than, or equal to, x. Since you have already solved the problem using $n = 1$, we will assume that $1 < n \leq 9$ and notice that
$$\lfloor \sqrt{\sqrt{n}} \rfloor = 1, \quad \lceil \sqrt{\sqrt{n}} \rceil = 2.$$

Also, we adopt the customary notation $\overline{nn}$ for the two-digit number $10n + n$ and, just as you did, write $.\bar{n} = .nnn\ldots = n/9$. Then we have the following representation:

$$1 = n^{n-n}, \quad 2 = \frac{n+n}{n}, \quad 3 = \sqrt{\frac{n-.n}{n}}, \quad 4 = \left\lceil \sqrt{\frac{\overline{nn}}{n}} \right\rceil, \quad 5 = \frac{n}{.n + .n},$$

$$6 = \left(\sqrt{\frac{n-.n}{.n}}\right)!, \quad 7 = \left(\sqrt{\frac{n}{.\bar{n}}}\right)! + \lfloor \sqrt{\sqrt{n}} \rfloor, \quad 8 = \frac{n - .\bar{n}}{.\bar{n}},$$

$$9 = \frac{n-.n}{.n}, \quad 10 = \frac{\sqrt{n \times n}}{.n}, \quad 11 = \frac{\overline{nn}}{n}, \quad 12 = \left\lceil \frac{n + .\bar{n}}{.n} \right\rceil.$$

This representation is not unique and, for explicit values of n, can be simplified substantially. As you see, mathematical 'grammar' is very sophisticated and includes some really powerful structural elements.

STU: The floor and the ceiling! We forgot about those two functions. But why eliminate 0 from the list of admissible digits? Can we not replace 1 by 0! on our clock and write $\overline{(0!)(0!)}$ for 11?

GAN: Well, Stu, you may have got something there.... Since you seem keen on this kind of exercise, here is an even bigger challenge: show that the problem can also be solved with any digit n used exactly *twice*.

STU: Only twice?!

GAN: Yeap. After which, if you really want to test yourselves, demonstrate that one *single* use of any digit $n \geq 3$ can reproduce all the clock face numerals. But I must warn you: the chains of unary operations you will need to put in place for that will look quite horrendous!

JJ switched off the VOTSIT and leaned back in his chair. "Now," he said, "it's time for your stories."

I finished the last of the coffee and pushed the empty cup aside. "OK, here we go. Today's brace don't have a mathematical flavor, but they relate to what we've been discussing. First one coming up."

A European visiting Sydney, Australia, stops his car at a corner where two locals are quietly watching the traffic.

"*Sprechen Sie Deutsch?*"[22] he asks.

The men look at him and say nothing.

"*Parlez-vous français?*"[23]

The Aussies do not blink an eyelid.

"*Parlate italiano?*"[24]

Not a muscle moves on the men's faces.

"*¿Hablan español?*"[25]

---

[22] Do you speak German? (German)
[23] Do you speak French? (French)
[24] Do you speak Italian? (Italian)
[25] Do you speak Spanish? (Spanish)

Still getting no response, the tourist guns the car engine and takes off, muttering indistinctly under his breath.

One of the Aussies turns to the other and says, "D'you think, you and me, maybe we should learn a foreign language?"

"Why?" the second one says. "That guy knew four, and a fat lot of good it did to him."

"Are you sure this happened in Australia and not closer to home?" JJ asked.

"Well, that's how I heard it. The incident in the second story, though, could've happened anywhere."

Two mice, father and son, are scurrying together along a wall, when, suddenly, a fierce cat pounces on them. While the son tries to blend into the wall stones, the father mouse opens his snout wide and bellows, "Woof-woof!!"

The cat draws up short, turns, and lopes away.

"You see, my son?" the father says to the young one. "Do you understand now why it's essential to learn foreign languages?"

"How very true. How very apt." JJ put his hands on the table. "Thanks for the stories," he said. "I've really enjoyed our little chat. If all goes well, we'll talk some more next time." Then, unceremoniously, he sprang to his feet and promptly walked out of the bar.

Notes After the Meeting

Here are the answers to JJ's questions together with some brief personal comments.

Solution to Question 1. Equality (6.2) is invalid. To keep things simple, we remark that the equation $z^2 = -1$ has two roots, namely, $\pm i$, so (6.2) should be replaced by

$$\frac{ci}{1} = \pm \frac{c}{i}.$$

The '+' sign leads to the anomaly in the 'proof', and must be discarded. After cross-multiplication, the '−' sign yields

$$-ci^2 = c,$$

which is the same as the obvious equality $c = c$. Consequently, the '−' sign is the legitimate one.

A more sophisticated explanation of the error involves the fact that the index laws are generally invalid for complex numbers.

All told, the Inuit, it seems, will not be drinking piña colada and sunbathing on their porches any time soon.

Solution to Question 2 (arithmetic). The fact that John has as many brothers as sisters can be illustrated by the simple diagram

where the line at the bottom represents the 'zero' level. The fact that Mary has twice as many brothers as sisters is illustrated by a modified diagram, where the two sides A and B are out of balance, with A twice as high as B:

```
                all boys
        A  ─────────────────
                              all girls less 1
                         ─────────────────  B


        ────────────────────────────────
```

The imbalance is caused by the addition of one element (John) to group A and the removal of one element (Mary) from group B. Hence, the difference between the levels of A and B is 2 units, as is the difference between the level of B and the zero level. This means that the family has 4 boys and 3 girls.

Solution to Question 2 (algebraic). Let x and y be the numbers of boys and girls in the family. Counting from John's point of view and then from Mary's, we can write
$$x - 1 = y,$$
$$y - 1 = \tfrac{1}{2} x.$$

This is a simple system of equations for x and y which can be solved, for example, by substituting $y = x - 1$ in the second equation. Its solution is
$$x = 4, \quad y = 3.$$

A Word to the Wise

Complex Numbers

A complex number has the general Cartesian[26] form
$$z = a + ib,$$

where a and b are real numbers called the real part and the imaginary part of z, respectively, and $i^2 = -1$.

---

[26] Cartesius is the Latinization of the name of René Descartes, French mathematician and philosopher, 1596–1650.

The conjugate $\bar{z}$ and the modulus $|z|$ of z are

$$\bar{z} = a - ib,$$
$$|z| = \sqrt{a^2 + b^2} = |\bar{z}|.$$

Addition and multiplication of complex numbers is performed as you might expect:

$$(a + ib) + (c + id) = (a + c) + i(b + d),$$
$$(a + ib)(c + id) = ac + ibc + iad + i^2bd = (ac - bd) + i(bc + ad).$$

Division and other fancy operations with complex numbers can be found in books on this topic.

A complex number z can also be written in polar form:

$$z = |z|e^{i\theta},$$

where $\theta = \arg z$, $-\pi < \theta \leq \pi$, is the argument of z, defined by

$$\theta = \begin{cases} \tan^{-1}(b/a) & \text{if } a > 0, \\ \tan^{-1}(b/a) + \pi & \text{if } a < 0,\ b \geq 0, \\ \tan^{-1}(b/a) - \pi & \text{if } a < 0,\ b < 0, \\ \pi/2 & \text{if } a = 0,\ b > 0, \\ -\pi/2 & \text{if } a = 0,\ b < 0, \\ \text{undefined} & \text{if } a = b = 0. \end{cases}$$

The Cartesian form is obtained from the polar form by means of the equalities

$$a = |z|\cos\theta, \quad b = |z|\sin\theta.$$

An important formula, named after Euler,[27] used in work involving complex functions is

$$e^{i\theta} = \cos\theta + i\sin\theta.$$

---

[27] Leonhard Paul Euler, Swiss mathematician and physicist, 1707–1783.

SCAM 7

Foreign Countries and Foreigners

> *Comment voulez-vous gouverner un pays où il existe 365 variétés de fromage?*[1]
>
> Charles de Gaulle[2]

> *America has a history of political isolation and economic self-sufficiency; its citizens have tended to regard the rest of the world as a disaster area from which lucky or pushy people emigrate to the Promised Land. Alternatively, they think of other nations as mere showplaces for picturesque scenery, odd flora and fauna and quaint artifacts.*
>
> Alison Lurie[3]

I met JJ again at a mathematics conference sponsored by Georgetown University in Washington, D.C. Just like the first time, he was sitting alone at a table in a corner of the hotel bar, gazing at me across the room with dark, inscrutable eyes.

As soon as I sat down, he pushed a piece of paper in my direction. "Here," he said laconically. "For your students."

I glanced at the paper and smiled. The text read something like this.

Question 1 (requires common sense). *The growth rate of a water plant on a pond is such that the total area covered by the plant doubles every week. If the entire pond is fully covered after exactly 10 weeks, after how many weeks had the plant covered half of the pond?*

Question 2 (requires basic arithmetic). *A long, long time ago, the inventor of the game of chess went to his country's Ruler and demonstrated it to him. The Ruler liked the new game so much, that he immediately offered to buy it and asked the inventor to name a price. "It's not cheap," the inventor said. "It took me many years to come up with this idea. But you can pay me in installments. Today, give me just one grain of wheat. Tomorrow, give me two grains. The day after, give me four grains, and so on, doubling the number of grains every day for 64 consecutive days, which is the number of squares on the chessboard. If you don't have enough wheat, then you pay me its market value in gold."*

---

[1] How can you govern a country that has 365 kinds of cheese? (French)
[2] Former president of France, 1890–1970.
[3] American author and educator, b1926.

The Ruler readily agreed with the inventor's price, believing he had a bargain. In his place, would you have done the same? If not, why not?

"Interesting," I said. "Thanks." I folded the paper and put it in my pocket, waiting for JJ to broach, as usual, a subject of his choice. He needed no special invitation.

"Some people are very parochial," he began abruptly. "They want to know only what's happening in their own backyard. For them, that's the entire world. Even everyday language gets colored by this view. Look at baseball: the last contest of the season is called the *world* series, although it is always between two American teams."

"You can't expect every JOE to know about everything."[4]

"No, but every JOE should be able to name at least three continents and tell where at least half a dozen countries are located. That this doesn't happen can be blamed on the poor quality of public school education, where, of all subjects, geography probably gets the most superficial treatment. What's worse, the gaps in the students' general knowledge at that level quite often tend to remain unfilled for the rest of their lives. Here is a sample collected by my colleague Gan from a popular TV quiz show, where the 'master of ceremonies', Mas, interrogates the contestant, Con. The questions and answers are from different episodes."

MAS: Which of the following countries is farthest east with respect to the Pacific Ocean? (a) Greece; (b) Thailand; (c) Syria; (d) Brazil.

CON: I'd like to ask the audience.

(The audience splits 29% for Greece, 25% for Thailand, 23% for Syria, and 23% for Brazil.)

CON: I think it's Greece, but I'm not sure. I'll stop and take the money.

MAS: What is the capital city of Belgium? (a) Dijon; (b) Milan; (c) Brussels; (d) Copenhagen.

CON: Mmm.... A tough one. Can I go 50-50, please?

(The computer reduces the choice to Brussels or Copenhagen.)

CON: They both sound right.... I don't want to risk, so I'll go for the money.

MAS: On what continent would you find the Zambezi river? (a) Europe; (b) North America; (c) South America; (d) Africa.

CON: I'd better phone a friend.

(The friend suggests South America, with a 25% degree of certainty.)

CON: I'm not convinced, so I guess I'll have the money.

"The disturbing part," JJ said, "is that all these contestants were professionals, supposed to have reached a creditable level of general education. If *they* have trouble with geography, then what can you expect from other JOEs? Recently, Gan amused himself with a survey for testing how a slight geographical

---

[4] Or, as Virgil (Publius Vergilius Maro, Latin poet, 70–19 BC) put it: *Non omnia possumus omnes.* (Not all of us are able to do all things.)

bias in the question may affect people's logic. 'What is 3 times 4 in Russia?' he asked. Most respondents were not fooled by the mention of a foreign country and gave the correct answer. But a few took the bait, making comments that ranged from perplexing to humorous. Here are some of them."

"Search me! They use metric there, don't they?"

"My calculator is American, not Russian."

"How should I know? Never been to Arkansas."

"It's what the Politburo[5] tells them it is."

"*Dvenadtsat'*."[6]

"Many people compensate for what they are not taught in school by reading on their own, traveling abroad, or meeting with natives of other lands," I said.

"True, but compared to the size of the population, that number is not large enough. Even individuals in places of higher learning manage once in a while to embarrass themselves pretty badly by vocalizing their lack of knowledge of foreign countries in situations where silence would be golden. Gan told me that a few of years ago his department had a visitor, Vis, from Eastern Europe, who gave a one-hour seminar on his research. After the talk, they all gathered in the faculty room for coffee: Gan, Vis, and a sprinkling of other academics, including a certain assistant professor, Ass. Part of the discussion went something like this."

GAN: What hobbies do you have, Dr. Vis?

VIS: I like to play chess, read thrillers, and go to the opera.

ASS (foolishly deciding to act in the spirit of his name): Opera? You've got opera in your country?

VIS (with perfect composure): I know it sounds incredible, but, yes, when we get fed up swinging from tree to tree, we put on our Sunday grass skirts and go to listen to Puccini.[7] Then, fully uplifted, we grab some wood stakes and dash into the forest to defang a vampire or two. These creatures have infested a wide area of the old continent and are becoming real pests. Besides, hunting the bloodsuckers brings a lot of pleasure to foreign tourists, who pay good money to join in.

ASS (trying to extract his foot from his mouth): Please excuse me, but I must go to a meeting.

"I bet Gan gets asked quite a lot what country he's from," I said.

"Frequently. When he has time, he challenges people to guess. There is no chance they might come up with Ganymede, of course, but he's curious to see how many country names they know. They all run out of options pretty quickly after going through a subset of a list featuring Canada, Mexico, United

---

[5] The communist policymaking body in the former Soviet Union.

[6] Twelve. (Russian)

[7] Giacomo Antonio Domenico Michele Secondo Maria Puccini, Italian composer, 1858–1924.

Kingdom, France, Germany, Spain, and Russia. On occasion, someone bold might mention Italy, Sweden, Portugal, or Poland, but that's about it. Once, though, Gan was really startled by a student with 'Sir, you must be from outer space.' 'Why?' Gan asked cautiously. 'It's your lectures,' the answer came. 'They are out of this world.' The student in question is now a successful politician."

"Since you mentioned students, does Gan ever bring geography into the classroom?"

"Certainly. Last semester, he set a simple encryption problem for a bunch of juniors as an illustration of the use of numbers in coding. What country, he asked, would they identify from the label 1140341? The solution was found by Stu, who volunteered to step up to the board and wield the marker. Here is his VOTSITed presentation."

STU: You said this was a *simple* problem, Dr. Gan, so I'll start with the simplest assumption, whereby each letter is assigned a natural number from 1 to 26, according to the position of the letter in the alphabet. (Writes on the board.)

A B C D E F G H I J K L M N O P Q R S T U V W X Y Z
1 2 3 4 5 6 7 8 9 10 11 12 13 14 15 16 17 18 19 20 21 22 23 24 25 26

Under my assumption, 0 does not designate a letter, so either (i) numbers less than 10 are written in double-digit form with 0 as the first digit, or (ii) 0 is the second digit of a double-digit number. Clearly, (i) is not true because the code you gave us consists of an odd number of digits. At the same time, the only number satisfying (ii) is 40, which is not assigned to any letter. Hence, what I called my simplest assumption was wrong.

GAN: Well reasoned, Stu. What next?

STU: I'll now go one step higher on the difficulty ladder and assume that the coding has been done using residue classes modulo 5.[8]

GAN: Why?

STU: I notice that the highest digit in your code is 4, and 0 has a legitimate place in this type of arithmetic, representing all multiples of 5. So, let's divide all the numbers above by 5 and write the corresponding remainder under each letter.

A B C D E F G H I J K L M N O P Q R S T U V W X Y Z
1 2 3 4 0 1 2 3 4 0 1 2 3 4 0 1 2 3 4 0 1 2 3 4 0 1

There are three 1s in the code: the first, second, and last digits. According to our table, they may represent A, F, K, P, U, or Z. The seven-letter country names I know that begin with one of these letters are Albania, Algeria, Andorra, Armenia, Austria, Finland, Ukraine, and Uruguay. Of them, Austria and Ukraine have the

---

[8] In plain language, the set of all possible remainders when an integer is divided by 5; that is, $\{0, 1, 2, 3, 4\}$.

second letter drawn from the same pool, but Ukraine fails to meet the last-letter criterion, whereas Austria satisfies it. If we check the remaining letters in the chart, we see that Austria is indeed the answer.

GAN: Excellent, Stu! You think logically, have insight, and also know about the countries of the world.

"I wish more people would travel abroad and get a glimpse of other nations' lives and customs," I said. "But this is not easy for Americans. While they can sneak into Canada or Mexico, particularly if they live near the border, a journey to Europe or Asia requires amounts of money and time JOEs can rarely afford."

"Enough JOEs manage it. That's not the problem. The real problem is the attitude and behavior of some of those who do visit foreign lands. The great majority of tourists, quite sensibly, expect certain cultural differences and try their best to accommodate them. A few, however, against all logic, show no such understanding and demand that the hosts provide for them the same kind of environment they have at home. Here is the gist of a conversation overheard by Gan and recorded on his VOTSIT, between a reception clerk, Rec, and an unpleasant twenty-something, Som, at a hotel in Provence."[9]

SOM: My room is too small and doesn't have air conditioning. Give me another one.

REC: We are in the mountains, sir; we don't need air conditioning. As for the rooms, they are all the same size. However, if it makes you happy, I could switch you to a different room. But you'll lose the beautiful view that Monsieur Cézanne[10] admired so much.

SOM: I don't care about the view, or about what your other guests think of it. And you call that a bathroom? With no shower? With a tub so small that it can barely hold one of my legs? The restaurant, too, is a disgrace: it serves no real food, only muck. How do you expect me to stay in this dump and pay for it? What kind of country is this?

REC (manages to keep calm): This is *my* country, sir. It happens that I like it a lot, just the way it is; with the muck we eat in our restaurants, which, if you ate it yourself, would help you lose 50 kg[11] within six months; with our small bathtubs, where, if you lost all that flab, you would find space enough for your other leg as well, and the rest of your body; with rooms that we ventilate by opening windows to let in the fresh, scented air of the Alps; and with views that have inspired some of the greatest artists in recent times. I can advise you that there is a train to Paris in about an hour, and then a transatlantic flight at 6 in the evening, which will take you back to where you can eat two dozen hamburgers in a hotel room the size of a tennis court while shivering under lashes of polar-cold air and looking at

---

[9] A picturesque region of southeast France, stretching between the Mediterranean Sea, the Alps, and the Rhône river.

[10] Paul Cézanne, French post-impressionist painter, 1839–1906.

[11] About 110 pounds, for the metrically illiterate.

a skyline that would make a normal person want to jump out the window—if only the windows could be opened. To each his own, sir, I suppose.

SOM (livid): Who do you think you are talking to? You miserable Frogs would have no country if we had not liberated it for you!

REC (maintains his calm): First, it's 'whom', not 'who'. I won't bother explaining to you why, because, after all, English is your native language. Second, what's this 'we'? Years ago, many brave men from your older generation laid down their lives chasing the enemy out of France. We will forever be grateful to them and honor their memory. But that's got nothing to do with you. You resemble more a specimen of a later generation, long on hair, flowers, and drugs and short on cleanliness, who knew better than their elders and burned the country's flag in public to prove it. You are just a rude, arrogant, whale-sized young lout with no manners and no sense of value. You bring disgrace on your countrymen who, unlike you, tour the world with decency and respect for the places and people they visit. You are an anomaly. I'm afraid that tonight there is no vacancy in this hotel for you, sir.

"According to Gan," JJ said, "Som could not find a room anywhere else in town, so he was forced to sleep on the grass in the local park, caught pneumonia, had to cut his trip short and flew back home—but he still thought he had won the contest with Rec."

"How did Gan learn those final details?"

"He happened to sit by Som on the plane back from Paris, and got an earful from him about those 'damn foreigners'. When Gan pointed out that the French were not foreigners in France, Som turned on him and accused him of being unpatriotic. Just before landing, Gan asked the young man if he saw anything wrong with the equality $a/(b+c) = (a/b) + (a/c)$, to which Som proudly replied that, as a high school graduate, he could confirm its validity. By way of a parting shot, Gan told him that he had just torn to pieces the Third Mathematical Commandment and that, as a consequence, he was an unBRITE with a MICQ no higher than 14."

"A highly atypical case," I said. "I haven't met anyone like Som on my travels abroad."

"You've been lucky. Then again, you go mostly to conferences, where the company is rather different."

"I can see how a few bad apples might make foreigners speak ill of us. But that's not the whole story, is it?"

JJ shook his head. "No, it isn't. Let's hear what a survey conducted by Gan outside the U.S. turned up on this subject."

GAN: What do you dislike about the Americans?

JOE: They believe they are the most powerful, militarily and economically, country in the world; they think they are entitled to chase and punish their enemies wherever they hide; they say that individuals should take full responsibility for their own lives; they claim that theirs is the land of opportunity, where anyone prepared to work hard has a chance to become very successful.

GAN: Anything specific that irritated you there?

JOE: Workaholism; obsession with litigation; exorbitant medical insurance costs; expensive medicines; exaggerated political correctness; too many trashy TV programs; sugar in almost every food; overhead electrical wires; reliance on volunteers to do what should be proper paid jobs; the basketball poles in front of nearly every house—how long do you want me to go on?

GAN: Anything specific that you liked?

JOE: The vast expanse of land; extensive highway network; ample parking; comfortable houses; quality medical care; availability of employment; low taxes; reduced union power; great community spirit; high living standard; all-you-can-eat places...

GAN: Would you like to live in America?

JOE: Yes.

GAN: Why?

JOE: Because it is, militarily and economically, the most powerful country in the world; it chases and punishes its enemies wherever they hide; it encourages the individuals to take responsibility for their own lives; it is the land of opportunity, where anyone prepared to work hard has a chance to become very successful.

"There you have it: the foreigners' love–hate relationship with your country." JJ leaned back in his chair. "Now," he said, "it's time for your stories."

I finished the last of the coffee and pushed the empty cup aside. "OK, here we go. First one coming up."

A mathematician goes on vacation to the Holy Land with his wife and mother-in-law. While there, the mother-in-law has a heart attack and dies. The man and his wife go to an undertaker, who explains that they can do one of two things: they can ship the body home, at a cost in excess of $3,000, or have her buried in the Holy Land, with a beautiful service, for only $300.

Almost immediately, the man says, "There is no question of a burial here. We'll ship her home."

Surprised, the undertaker says, "But you are a mathematician—you understand the great difference between the two figures. Why not opt for the more economical version? Is there something in her will about where she wanted to be buried?"

"This has nothing to do with the figures, or her will," the man answers. "Don't you remember? Two thousand years ago they buried a guy here and three days later he rose from the dead. With my mother-in-law I just can't take that kind of chance, so home she goes."

"Numbers and a foreign land: not bad," JJ noted. "You have a second one?"

"I do. More foreigners and numbers."

At the Austrian–Italian border, an Italian guard stops an incoming Austrian-registered car and steps up to the driver's window. Since neither one of them speaks the other's language, they both turn to English.

"Vot is wrong, officer?" the driver asks.

"You are-a five-a *persone* in an Audi Quattro. One-a *persona* too many."

"Zis is a car for five people," the driver replies. "I got ze documents to prove it."

"'Quattro' means-a 'four' in Italian. Issa not *permesso* to have-a more than-a four people in dis car."

The Austrian driver loses his cool. "Vot are you talking about? Zer is no such regulation."

"*Mi spiace*,[12] but I have-a my orders."

"You are a stupit man! I vont to speak viz somebody in authority, somebody viz more *intelligenz*. Ver is your zupervisor?"

"You cannot-a speak wiz-a him right now," the guard replies. "He's sorting out two men-a in a Fiat Uno."[13]

"And here's a third one," I said, "which also tries to illustrate the two related but different profiles of our discipline."

A pure mathematician and an applied mathematician, traveling in a faraway country, one day fall foul of the law and get hauled before the country's Ruler.

"You have behaved despicably," the Ruler says to them. "You have been caught selling reformed calculus books to our students, which in this land is a criminal offense. You must, therefore, suffer the full rigor of the law, as an example for others who might feel tempted to try to pervert the minds of our young with short-lived fashionable trash.[14] But since you are mathematicians, and since we pride ourselves on being a civilized people, I'm prepared to show you all the leniency that our customs allow."

The Ruler looks at them with fierce, linearly independent eyes, takes them to an empty room and explains, "You will be placed with your backs against the near wall, hands tied behind you. Every day, you will be allowed to advance half of the distance between where you are and the far wall. You will be forced to stand up straight at all times and will be given water but no food until you reach the far wall, which is coated with chocolate. If you decide to go through this and survive, you will be freed. If you fail, you will die a particularly horrible death. If you decline to take the challenge, then I will spare you and send you to prison for life. So, what's it to be?"

The pure mathematician has no hesitation and whispers to his partner in crime, "I choose life imprisonment."

---

[12] I am sorry. (Italian)

[13] Fiat Uno is a model of the most popular Italian car. Fiat is an acronym: *Fabbrica Italiana Automobili Torino* (the Italian Automobile Factory Torino). In Italian, *uno* means 'one'. It seems that Audi, too, is an acronym, although it didn't start that way. Audi is the Latin translation of the imperative mood of the old German verb *horchen* (to listen), which was the name of the first manufacturer of the car. Later, the company amalgamated with three others and the new entity decided on a name to fit Audi as an acronym: *Auto Union Deutscher Industrie* (the Automobile Consortium of German Industry).

[14] Gan will be encouraged to hear that there are still countries where reformed calculus is seen for what it really is.

"Are you crazy?" the applied mathematician whispers back. "Why spend the rest of your days behind bars in this soul-destroying place?"

"But don't you see? The numbers are not on our side. This is a classical problem involving a geometric progression with common ratio $1/2$. Let L be the length of the room. After the first day, I'll have advanced a distance of $\frac{1}{2}L$ toward the far wall; after the second one, I'll have advanced another $\frac{1}{4}L$; after the third one, another $\frac{1}{8}L$, and so on. If I sum up all these distances for *infinitely many* days, I get

$$\tfrac{1}{2}L + \tfrac{1}{4}L + \tfrac{1}{8}L + \cdots = L\big(\tfrac{1}{2} + \tfrac{1}{4} + \tfrac{1}{8} + \cdots\big) = L\,\frac{\tfrac{1}{2}}{1-\tfrac{1}{2}} = L,$$

which is the total distance I need to cover to reach the chocolate wall. This means that I'll never get there. So I'll save my life and go to prison instead. It won't be too bad, you know—it'll give me time to prove some nice theorems about compact operators on Banach[15] spaces that I've been thinking about for a while. I suggest you do the same."

"No way!" the applied mathematician exclaims, sweeping his arm widely. "What's the length of this room? I'd say 10 m tops. If I put $L = 10$ in the model and compute the distance that I'm allowed to cover in 8 days, I come up with

$$\big(\tfrac{1}{2} \times 10\big) + \big(\tfrac{1}{4} \times 10\big) + \big(\tfrac{1}{8} \times 10\big) + \big(\tfrac{1}{16} \times 10\big) + \big(\tfrac{1}{32} \times 10\big)$$
$$+ \big(\tfrac{1}{64} \times 10\big) + \big(\tfrac{1}{128} \times 10\big) + \big(\tfrac{1}{256} \times 10\big),$$

which is roughly 9.96 m. Therefore, after 8 days I'll be about 4 cm away from the far wall." He sticks out his tongue.

"No need to be rude," the other man says.

"I'm not being rude," the applied mathematician explains. "I'm showing you that my tongue is at least 5 cm long, so after 8 days I'll be able to lick that life-saving chocolate and go free."

"A good point concerning the accuracy of an approximation. Pragmatism over purity." JJ put his hands on the table. "Thanks for the stories," he said. "I've really enjoyed our little chat. If all goes well, we'll talk some more next time." Then, unceremoniously, he sprang to his feet and promptly walked out of the bar.

Notes After the Meeting

Here are the answers to JJ's questions together with some brief personal comments.

Solution to Question 1. The key word is 'doubles'. Going backwards in time from the end of the tenth week, we can say that each week the area covered is

---

[15] Stefan Banach: Polish mathematician, 1892–1945.

halved. Since the entire area of the pond is covered after 10 weeks, we conclude that at the end of the ninth week exactly half of it was covered. Therefore, the answer is 9 weeks.

Although this is a primary school question, many adults stutter and stumble before answering it. The split is about 1:6 between the correct solution and the 'preferred' wrong one of 5 weeks.

It all boils down to understanding numerical progressions. For the reader who had algebra some time ago, an *arithmetic* progression is a sequence where the *difference* between any number and the preceding one is the same; for example, 2, 4, 6, 8, ... In a *geometric* progression, it is the *ratio* of any number to the preceding one that is the same; for example, 2, 4, 8, 16, . . . , where, as in our problem, the numbers double at every step.

It seems that the human brain is naturally wired to cope with arithmetic progressions as the default option, not with geometric ones. This is the only plausible explanation (apart from, perhaps, ignorance and greed) for the woefully wrong decision made by the buyer in the second problem.

Solution to Question 2. The Ruler, who was very rich but not sufficiently bright and did not have a pocket calculator to do the arithmetic for him, or a mathematical advisor, thought that this was a price he could more than amply cover with a couple of bags of wheat, so he drew up and signed a bill of sale on the inventor's terms. A terrible error, he soon found out, as a simple computation shows.

The chess table has 64 squares. If, for convenience, we use powers of 2 to write the number of grains due to the inventor each day for 64 days, then the total number comes to

$$1 + 2 + 2^2 + 2^3 + 2^4 + \cdots + 2^{63} = \frac{2^{63} \times 2 - 1}{2 - 1} = 2^{64} - 1$$

$$= 18,446,744,073,709,551,615.$$

Since a grain of wheat weighs about 0.75 gram, the Ruler had to give the inventor some 13.8 trillions of tonnes of wheat—a mind-numbing amount.

Legend has it that, not long after, this incautious Ruler was forced to sell everything he owned in order to pay the inventor as much of the debt as possible. The Ruler finished his life in abject poverty, and, to make matters worse, he never became a proficient chess player either.

A Word to the Wise

Algebraic Operations

For simplicity, in what follows, the word 'brackets' denotes ( ) (parentheses) or [ ] (brackets) or { } (braces).

☺ DO

(i) Use brackets to separate successive arithmetic operation symbols:
$$a - (-b) = a + b,$$
$$a \times (-b) = a(-b) = -ab.$$

(ii) Remember that every term in brackets is affected by operation signs such as $-$ or $\times$ (whether explicit or implicit) immediately preceding the brackets:
$$a - (3b - 2c) = a - 3b + 2c,$$
$$a(3b - 2c) = 3ab - 2ac.$$

☠ DON'T

Here is something that, if written, might damage one's mathematical health:

☠ $\quad a - -b, \quad a \times -b,$

☠ $\quad a - (3b - 2c) = a - 3b - 2c,$

☠ $\quad a(3b - 2c) = 3ab - 2c,$

☠ $\quad -(ab) = (-a)(-b).$

☺ DO

Algebraic operations in any complex expression must be performed according to the following rule of precedence:

(i) brackets;
(ii) powers, roots;
(iii) multiplications, divisions;
(iv) additions, subtractions.

Here are a few examples:
$$15 - 4 \times 3 = 15 - 12 = 3,$$
$$(15 - 4) \times 3 = 11 \times 3 = 33,$$
$$8 \div 4 + 3 \times 5 = 2 + 15 = 17,$$
$$(2 + 3) \times (5 - 1) = 5 \times 4 = 20,$$
$$17 - 3 \times (4 - 2)^2 = 17 - 3 \times 2^2$$
$$= 17 - 3 \times 4 = 17 - 12 = 5.$$

In a chain of several consecutive operations of the same class, the operations are performed in the order in which they occur. Thus,
$$3 \times 4 \div 6 = 12 \div 6 = 2,$$
$$8 - 6 + 3 = 2 + 3 = 5.$$

A caveat: the powers referred to in (ii) above must be single numbers; if they include other operations, then those operations must be performed first:

$$2^{3^2} = 2^{(3^2)} = 2^9 = 512.$$

☠ DON'T

It is wrong to ignore the rules of precedence and 'simplify' things by performing all operations in the order they are written:

- ☠ $15 - 4 \times 3 = 11 \times 3 = 33,$
- ☠ $8 \div 4 + 3 \times 5 = 2 + 3 \times 5 = 5 \times 5 = 25,$
- ☠ $2^{3^2} = (2^3)^2 = 8^2 = 64.$

Life can become very complicated when such errors are committed.

SCAM 8

Mathematics and the Public

> *Timendi causa est nescire.*[1]
> Lucius Annaeus Seneca[2]
>
> *I have often admired the mystical ways of Pythagoras,*[3] *and the secret magic of numbers.*
> Sir Thomas Browne[4]

I met JJ again at a mathematics conference sponsored by Harvard University in Cambridge, Massachusetts. Just like the first time, he was sitting alone at a table in a corner of the hotel bar, gazing at me across the room with dark, inscrutable eyes.

As soon as I sat down, he pushed a piece of paper in my direction. "Here," he said laconically. "For your students."

I glanced at the paper and smiled. The text read something like this.

Question 1 (requires insight). *There are 14,539 competitors taking part in a single-elimination chess tournament. How many games need to be played to find the winner? (In the case of a stalemate or a draw, the winner is decided by the toss of a coin.)*

Question 2 (requires elementary logic). *A man M, a woman W, and a child C are facing together a table on which lie two sealed, identical-looking envelopes. They are told that there is money in both envelopes, but that one of them contains twice as much money as the other. M and W are invited to pick up one of the envelopes. As they do so, they are advised that, should they wish, they could exchange it for the one on the table, and that, whichever envelope they choose in the end, they can keep the money in it.*

M says, "We should hold on to what we have now, because if we give up this envelope for the one on the table, we stand to lose twice as much as we stand to gain."

"How do you figure that?" W asks.

"Well, let s_h be the sum inside the envelope in our hands, and let s_t be the

---

[1] Ignorance is the cause of fear. (Latin)
[2] See footnote on p. 21.
[3] Greek philosopher and mathematician, c582–c500 BC.
[4] English author, 1605–1682.

sum in the envelope on the table. If $s_h > s_t$, then

$$s_h = 2s_t,$$

so the loss by exchange would be

$$s_h - s_t = 2s_t - s_t = s_t.$$

On the other hand, if $s_h < s_t$, then

$$s_h = \tfrac{1}{2} s_t,$$

so the gain by exchange would be

$$s_t - s_h = s_t - \tfrac{1}{2} s_t = \tfrac{1}{2} s_t,$$

which is half of the potential loss."

"No, no," W says, "you got the whole thing the wrong way around. If we make the exchange, then we stand to gain twice as much as we stand to lose. See, if $s_h > s_t$, then

$$s_t = \tfrac{1}{2} s_h,$$

so the loss by exchange would be

$$s_h - s_t = s_h - \tfrac{1}{2} s_h = \tfrac{1}{2} s_h.$$

If $s_h < s_t$, then

$$s_t = 2s_h,$$

so the gain by exchange would be

$$s_t - s_h = 2s_h - s_h = s_h,$$

which is twice as much as the potential loss."

At this point, C looks at M and W and says, "You are both wrong. If you exchange the envelope you now have in your hands for the one on the table, you stand to gain exactly as much as you stand to lose."

"Why?" M and W chime in unison.

"It's obvious. Let n be the smaller of the two amounts s_h and s_t. If $s_h > s_t$, then

$$s_h = 2n, \quad s_t = n,$$

so the loss by exchange would be

$$s_h - s_t = 2n - n = n.$$

If $s_h < s_t$, then
$$s_h = n, \quad s_t = 2n,$$
so the gain by exchange would be
$$s_t - s_h = 2n - n = n,$$
which is exactly the same as the potential loss."

Now who is right: the financially pessimistic M, the gambling-inclined W, or the private-school educated C?

"Interesting," I said. "Thanks." I folded the paper and put it in my pocket, waiting for JJ to broach, as usual, a subject of his choice. He needed no special invitation.

"If you think about it," he began abruptly, "the public at large displays a very unhealthy attitude toward mathematics. Many proclaim loudly that they have never been able to cope with numbers, regarding their lack of numeracy as a badge of honor. How can one be proud of one's ignorance?"

"They probably think of mathematics as a devilish, unnatural pursuit and want to distance themselves from it."

"Even some educated people, astonishingly enough, allow their lack of elementary mathematical knowledge to taint their work, language, and general thinking. Highly illogical. As you might expect, our team has compiled a long list of classical howlers."

I waited for JJ to tell me, and he obliged.

"Remember what the great majority of the populace thought they were celebrating on December 31, 1999?"

"The end of the second millennium," I said with a smile.

"Indeed! Could those unBRITEs not count? Year 1 ran from January 1, 0001, until December 31, 0001. Year 2 ran from January 1, 0002, until December 31, 0002. Year 3 ran from January 1, 0003, until December 31, 0003. Following this pattern, they should have seen that Year 1999 began on January 1, 1999, and would finish on December 31, 1999, and that Year 2000, the last year of the second millennium, would run from January 1, 2000, until December 31, 2000. Therefore, the second millennium was to be completed at midnight on December 31, 2000. For a long time even people in the media, whom you would expect to be better informed, made the same error."

"Yes, but then they quietly changed their tune, saying that the planned worldwide celebrations on December 31, 1999 were, in fact, organized to mark the advent of the Year 2000."

"That happened very late in the game, when someone must've told them that they were being dense," JJ said. "And what about the sport performers interviewed on radio and television who promise to give 110% in next day's competition? Don't they know that any entity in the universe consists of only 100% of itself? Where do they think the extra 10% will come from?"

"Well, it's just a figure of speech. What they mean is that they will give everything they've got."

"But that still makes it a full 100% and no more. Where were they when their teachers taught percentages in school? Speaking of percentages, my colleague Gan recently witnessed an argument between a sales associate, Sal, and a confused shopper, Con, about the price of a garment that had been reduced twice. Here is what he recorded on his VOTSIT.

CON (points to a large overhead banner proclaiming that all items are subject to a 50% price reduction, and then to another one written in red saying that every item was discounted that day by a further 50%): Why do you ask me to pay for it? The original price of this blouse was $80. Discounting 50% twice means a total reduction of 100%, so I should have this blouse for free.

SAL: Ma'am, the register says that you have to pay $20.

CON: The register is wrong. The new price should be $0. Don't you agree that $50 + 50 = 100$? What did they teach you at school?

SAL: You see, the first reduction cuts the price in half, from $80 to $40, and the second one cuts the new price in half again, bringing it to $20. This leaves a balance of $20.

CON: This leaves a balance of nothing.

SAL: What if the red sign said that the extra reduction was 70%? According to you, that would make a total discount of 120%. Do you think the shop should give you the blouse *and* an additional $16 in cash?

CON (with supreme confidence): Don't be silly. You know nothing about elementary arithmetic. Call the manager.

"The vociferous shopper turned to Gan as if expecting confirmation that she was right," JJ said.

"So what did Gan do? Lecture her on compounding percentages?"

"It would've been a total waste of time. The woman was an unBRITE. Gan bit his tongue and walked away from the line."

"Some stores are now trying to prevent such occurrences by making it clear that the final price is 50% of the previously reduced price."

JJ shook his head. "It might alleviate the problem, but it won't cure it. Not when innumeracy is so widespread. A few months ago, I read on the internet an article by a college-educated writer who made two laughable statements. First, she claimed that, *by definition,* half of the population is below average in mathematical ability. She obviously doesn't understand the concept of average. In a population of 4 with test scores of 1, 5, 5, and 5, the average score is $(1 + 5 + 5 + 5) \div 4 = 4$, yet only one individual has a figure below that."

"She must have confused the arithmetic mean with the median," I observed.

"Second," JJ said, "she claimed that someone who has mostly grades of A and a grade of C can still have a 4.0 grade point average. A grade point average of 4.0 is obtained when the student has only grades of A. Even in our colony of educational delinquents on Callisto the level of mathematics is better than that."

"Not everybody is endowed with a liking for, and understanding of, numbers."

"That aside," JJ said, "such lack of elementary education is quite worrying. What's even more worrying is to see how frequent misuse of mathematical terms by the public at large gets absorbed and rooted into everyday language, which then is allowed to prevail over the precision and rigor of mathematics. Last year, Gan saw proof of this in a small claims court, where a customer, Cus, who had bought a TV set from an electronics store, was suing the manager, Man, of the store for deception and wanted his money back. The verbal exchanges between them and judge Jud went something like this."

JUD (to Cus): How much did you pay for the set?

CUS: $1,000, sir.

JUD: Was this the price marked on the ticket?

CUS: It was.

JUD: So what's wrong, then? You knew the price before you made the purchase.

CUS: The thing is, there was this big sign in the store saying that all TV sets were being offered at a fraction of the manufacturer's recommended price. A couple of days after I bought the set, I discovered that the MSRP was only $900. I had been deceived.

JUD (to Man): Why did you lie to him?

MAN: I didn't lie, sir.

JUD: How's that? The MSRP was $900. You claimed that the sale price was a fraction of the MSRP, yet you charged him $1,000, which is more than the MSRP. This sounds like lying to me.

MAN: I didn't lie, your honor. The price I charged him is $10/9$ of the MSRP, and, by all accounts, $10/9$ *is* a fraction. The sign in the shop did not say that the fraction applied to the MSRP was less than 1. Surely, a highly educated person like yourself would agree with this obvious mathematical truth.

JUD (to Man, suppressing a smile): I appreciate your defense argument more than you think, and your groveling. If this were a mathematical court, you would walk away from it a hands-down winner. Unfortunately for you, this is a court for the common people, and in common people's language, a fraction is taken to mean a portion of the whole. I therefore must find in favor of the plaintiff.

"I wish the judge had struck a blow for arithmetic," JJ said. "I wish he had dismissed the case."

"But Man was a deceitful crook. The judge was right."

"All the same, it's heartening to see that some ordinary folk have a good grip on the concept of fraction, whatever their morals. Crook or not, Man had certainly not wasted his time in his school math classes."

"So you think that mathematics always loses when confronted with the language of the man in the street?"

"Not at all," JJ said. "As a matter of fact, I have proof of this. I have proof that, when the case is properly presented, and when the judge is open-minded and fair, a mathematician can come out on top."

SCAM 8

Sipping at my coffee, I waited patiently for JJ to tell me the story.

"On a televised quiz show," he said, "a college math professor, Mat, stood to win a million bucks if he answered correctly one last question. By an uncanny coincidence, the question was about numbers. What number, it asked, was the next one in the sequence $2, 12, 36, \ldots$? Mat thought long and hard about it, then said that it could be any number. Annoyed, the show host pressed him to make a choice and give him a specific number, so Mat said '74'. To the consternation of the live audience, who had expected the professor to find the number easily, the host announced that the correct answer was 80. Mat disagreed and took the show producers to court for loss of earnings by reason of their stupidity. When they appeared before judge Jud, both parties—Mat, representing himself, and Att, the producers' attorney—asked that the case be heard without a jury. Gan was in court and captured the proceedings on his VOTSIT.

JUD (to Att): Would you explain to me, in layman's terms, why the correct answer to the question is 80?

ATT (switching on the computer connected to the court projector and pointing to the screen): Yes, sir. The numbers in the sequence are formed by adding the square and the cube of each natural number. Thus, we have

$$1^2 + 1^3 = 1 + 1 = 2,$$
$$2^2 + 2^3 = 4 + 8 = 12,$$
$$3^2 + 3^3 = 9 + 27 = 36,$$

therefore, the next one is

$$4^2 + 4^3 = 16 + 64 = 80.$$

I'm surprised that Dr. Mat, who uses numbers as his stock-in-trade, did not spot this pattern. (Walks back from the computer and sits down, looking pretty pleased with himself.)

JUD (to Mat): How do you respond to that? It seems to me a very clear and valid explanation.

MAT (goes to the computer and works its keys): With the court's permission, I wish to make several points. First, Mr. Att, whose knowledge in matters legal I do not dispute but whose grasp of elementary algebra I very much doubt, makes a huge error when he presumes that a mathematician's work consists in shuffling numbers. That's what an accountant does. By contrast, a true mathematician deals in abstract concepts and logical thinking on a plane far above the numerical realm, completely unfathomable to someone outside the profession. Second, the answer he is trying to push as being correct is only one of the infinitely many that can be given to the question. My answer is equally valid, because, in my view, the numbers in the sequence are the values of the quadratic polynomial

$$P(x) = 7x^2 - 11x + 6$$

for x equal to each of the positive integers.

JUD (whispers to his clerk): What's a quadratic polynomial? (Loudly, after the clerk shrugs): Can you recall for us the definition of a quadratic polynomial, Dr. Mat?

MAT (works the computer again): A quadratic polynomial, your honor, is an expression of the form

$$P(x) = ax^2 + bx + c,$$

where a, b, and c are given numbers and x is called a 'variable'. For each number x we feed into it, the expression produces another number, $P(x)$, called the value of P at x.

JUD: Fine. But why 'quadratic'?

MAT (weighs the pros and cons of giving unprofitable details): It's because there are too many polynomials around and we need some way of keeping tabs on them. In any case, my specific polynomial yields

$$P(1) = (7 \times 1^2) - (11 \times 1) + 6 = 7 - 11 + 6 = 2,$$
$$P(2) = (7 \times 2^2) - (11 \times 2) + 6 = 28 - 22 + 6 = 12,$$
$$P(3) = (7 \times 3^2) - (11 \times 3) + 6 = 63 - 33 + 6 = 36,$$
$$P(4) = (7 \times 4^2) - (11 \times 4) + 6 = 112 - 44 + 6 = 74.$$

The first three values are those given in the quiz question and the fourth one is my answer.

ATT (jumps to his feet): Objection!

JUD: On what basis?

ATT: Dr. Mat is trying to hoodwink us with things we don't understand.

MAT (to Jud): Your honor, Mr. Att's not understanding my explanation doesn't make it wrong. It only exposes his mathematical ignorance.

JUD (probably not wanting to have his own knowledge of elementary algebra called into question): Objection overruled. You may continue.

MAT: The numbers in the given sequence can also be generated by other polynomials, or, indeed, by other, more general, functions. Not only that, but we can find polynomials (to keep the matter simple) that can generate any number of our choice as the fourth term in the given sequence. For example,

$$P(x, n) = -\tfrac{1}{6}(74 - n)x^3 + (81 - n)x^2 - \tfrac{11}{6}(80 - n)x + (80 - n)$$

is such that

$$P(1, n) = 2, \quad P(2, n) = 12, \quad P(3, n) = 36, \quad P(4, n) = n.$$

This shows that the sequence Mr. Att expected is generated by

$$P(x, 80) = x^3 + x^2, \quad x = 1, 2, 3, 4, \ldots,$$

whereas mine is generated by

$$P(x, 74) = 7x^2 - 11x + 6, \quad x = 1, 2, 3, 4, \ldots.$$

If we want the next number to be, say, 0, then we would use

$$P(x,0) = -\tfrac{37}{3}x^3 + 81x^2 - \tfrac{440}{3}x + 80, \quad x = 1,2,3,4,\ldots.$$

In conclusion, the quiz question admits infinitely many correct answers, each of which is obtained by means of infinitely many different formulas. We, mathematicians, call such a question *ill-posed*.

JUD: But Mr. Att's answer appears to be the most natural of all.

MAT: It certainly does. And it would indeed have been the *only* answer if the question had been *well-posed*. The question should have read, 'What is *the most obvious* number that comes next in the sequence $2, 12, 36, \ldots$?' Then I would have answered 80 and got my million, without any need for introducing polynomials into the argument.

JUD: Anything else, Dr. Mat?

MAT: Yes, your honor. Third, the ignoramuses represented by Mr. Att publicly humiliated me on television in front of the whole country by making me, a professor of mathematics and doctor of philosophy, look like a nincompoop who cannot handle a few simple numbers. To atone for this affront and to force others to take mathematics seriously in real life, I respectfully ask the court to award me not only the million to which I feel I am entitled, but also punitive damages in the amount of a further three million dollars.

"After the trial," JJ said, "it transpired that Jud gave himself a crash course on polynomials and linear systems of algebraic equations before ruling in Mat's favor. Mat was thus vindicated. With hindsight, I'd say that Mat knew full well what the expected answer was, but, being a BRITE, came up with a different one on purpose, to extract an additional tidy sum from the producers as price for their lack of mathematical education. A risky strategy that paid off."

"This isn't true!" I exclaimed. "This never happened!"

"Why don't you believe me?"

"Give me your VOTSIT!"

"You know what they say in Rome: *se non è vero, è ben trovato*.[5] I knew you were the right person to choose for my conversations!" JJ leaned back in his chair. "Now," he said, "it's time for your stories."

I finished the last of the coffee and pushed the empty cup aside. "OK, here we go. First one coming up."

Theorem. To a computer scientist, Halloween and Christmas fall on the same day.

Proof. As is well known, Halloween is observed on October 31 and Christmas on December 25, or, in abbreviated form, Oct 31 and Dec 25. A computer scientist may interpret Oct 31 as the *octal* number 31_8, that is, 31 written in base 8. Translating this into a *decimal* number (one in base 10), we have

$$31_8 = 3 \times 8 + 1 = 25_{10};$$

---

[5] They also say it in the rest of Italy: even if it is not true, it is well conceived.

in other words, Oct 31 = Dec 25, which means that the two dates indeed coincide.[6]

"On a par," JJ commented, "with the computer scientists' jocular claim that there are only 10 kinds of people: those who understand binary numbers and those who don't."

"And here's the second story," I said.

> Little Tommy walks home from school.
> "How was your day?" his mother asks him.
> "Fine. But I've got a drinking problem."
> The mother is shocked. "Says who?"
> "The teacher."

"Well, son, we need to wait until your father comes back tonight. We'll discuss this with him, to see what we need to do to help you."

A couple of hours later, Tommy's father arrives and his wife tells him about Tommy's drinking problem. "How should we handle it?" the woman asks, wringing her hands.

"This is beyond our competence," the father says. "We must go immediately to a psychiatrist."

The following morning, all three of them get into the car and drive to the local clinic, where the shrink on duty receives them in his office. The parents tell the shrink the reason for their visit.

"Tommy," the psychiatrist says, turning toward the little boy, "will you describe this drinking problem in your own words? I want to hear it from you before we decide how to solve it."

"Yes, sir," Tommy replies calmly. "It's like this. Three men drink from three bottles of water. One drinks twice as fast as the second one, who drinks twice as fast as the third. If the third man finishes drinking his whole bottle in eight minutes, how long do the others take to finish theirs?"

"I wonder if the shrink knew the answer," JJ said. "If he'd come through the public school system, chances are he didn't."

"Finally, the third story of the day."

A mathematician and an engineer are having a drink in a bar, arguing about numeracy in the population at large. The mathematician maintains that the public school system is doing such a shoddy job that JOE barely has a few rudiments of arithmetic. The engineer disagrees, saying that his friend might be surprised to find out how many ordinary people have, in fact, mastered quite a bit of basic algebra and even some simple elements of calculus. The mathematician sneers, gets up, and goes to the restroom.

In his friend's absence, the engineer calls the young waitress and says to her, "I'll give you a good tip if you do me a favor. When my friend returns, I'll ask you

---

[6] The interpretation of Oct and Dec as abbreviations for octal and decimal is reasonable, since October and December were, respectively, the eighth and tenth months of the Roman calendar, named for the Latin numerals *octo* (eight) and *decem* (ten).

a question. You don't need to worry about what it means; just look at me and answer 'A half of x squared.' Can you do that?"

"Sure thing," the waitress agrees. "Is this all you want me to say? A half of x squared?"

"Yes. A half of x squared. That's all."

A few moments later, the mathematician is back. With a Mephistophelean[7] smile on his face, the engineer says, "I think you're wrong. I'm convinced that even our young waitress, to pick a random example, must know some higher mathematics. In fact, I'm so certain, I'm willing to put money on it."

"A bet? What kind of bet?"

"I'll ask her a question on integration. If she answers correctly, you pay me 100 dollars. If not, I pay you 100 dollars."

"A question on integration, you say. Umm..." The mathematician gives his friend a long look. "But her answer must be absolutely correct."

"Agreed."

"Okay, you're on."

The engineer calls the young woman over and asks her, "Tell me, please: what is the integral of x?"

"A half of x squared," she answers without hesitation.

"See? I win!" the engineer exclaims with glee. "You owe me 100 bucks."

"I don't think so," the mathematician says. "Her answer is incomplete."

The engineer seems nonplussed. "What are you talking about? That's the correct answer: the integral of x is a half of x squared!"

With a shrug, the waitress turns to him and chirps, "No, sir, it's not. When we spoke a few minutes ago, you should have told me to add 'plus C'." As she walks away, she looks back and smiles impishly: "You are an engineer, aren't you?"

"Wicked!" JJ put his hands on the table. "Thanks for the stories," he said. "I've really enjoyed our little chat. If all goes well, we'll talk some more next time." Then, unceremoniously, he sprang to his feet and promptly walked out of the bar.

Notes After the Meeting

Here are the answers to JJ's questions together with some brief personal comments.

Solution to Question 1. Each game has one and only one loser, and each competitor (except the winner of the tournament) loses once and only once. Consequently, we can uniquely associate each game with its loser and then count the total number of games played by counting the total number of losers, which is $14,539 - 1 = 14,538$.

---

[7] Mephistopheles is the devil in the tragic play *Faust* by Goethe (see footnote 4 on p. 55) and in the opera with the same title by Charles Gounod (French composer, 1818–1893).

This is another little gem that shows the importance of lateral thinking. Instead of clobbering the data to death, pairing off the players and keeping tabs on the number of games in each eliminatory round, we take a step back and look for a simpler way to get to the core of the problem. The correspondence mentioned in the solution, between the elements of the set of games and those of the set of losers, is called a *bijection*. If you don't like this name because you think of it as one of those pompous-sounding words that mathematicians use to impress the general public, then you may also call it, quite safely, a 'one-to-one and onto' correspondence. How important bijections are and how we sometimes use them in real life without giving them a second thought can be seen from a simple example. When you look at a book and want to know how many pages it has, you don't start counting them—you just read the number printed on the last one (and make an allowance for the front matter). This is because each page has one and only one number, assigned consecutively, and each number is assigned to one and only one page.

Try this question on any number of people you like, and I'm almost certain that none of them will see the direct solution. They will all clobber the data to death.

Solution to Question 2. With no additional information, each of the three proofs seems to be logically correct. The conclusions are different because M, W, and C use different 'yardsticks' in their reasoning: M uses s_t, W uses s_h, and C uses $\min\{s_h, s_t\}$. As given, the problem is ill-posed, that is, its formulation is too vague, lacking a firm 'anchor' for the data. If we moore it to reality by saying or assuming something like 'let s be the total sum split between the two envelopes', then things become clear: the two individual sums are $\frac{1}{3}s$ and $\frac{2}{3}s$, and young C's argument is the correct one.

The argument is independent of whether or not M and W are married to each other, whether or not C is their child, and whether C is male or female. The cast of characters was left ill-posed to mirror the nature of the problem itself.

A Word to the Wise

Writing Fractions

Every nonzero number x has a reciprocal $1/x$, also written as x^{-1}. The number 0 does not have a reciprocal because division by zero, as already mentioned, is undefined. Then for any $x \neq 0$ and any a,

$$a \div x = a \times x^{-1} = a \times \frac{1}{x} = \frac{a}{x}.$$

 DO

When writing a fraction within a fraction, make sure that the main fraction line is slightly longer (which identifies it more easily) and level with the arithmetic operation signs in the formula. Note the difference between the values of the following two expressions:

$$\frac{\frac{a}{b}}{c} = \frac{a}{b} \div c = \frac{a}{bc},$$

$$\frac{a}{\frac{b}{c}} = a \div \frac{b}{c} = a \times \frac{c}{b} = \frac{ac}{b}.$$

 DON'T

The expression

$$\frac{a \times d}{\frac{b}{c}}$$

is ambiguous.[8] The main fraction line cannot be identified, and the operation of multiplication by d is misaligned. This type of writing is sloppy in the extreme and leads to a lot of computational errors in tests and homework.

---

[8] And an eyesore to boot.

SCAM 9

The Question of Calculus

> *A great licentiousness treads on the heels of a reformation.*
> Ralph Waldo Emerson[1]

> *I own I never really warmed*
> *To the reformer or reformed.*
> Robert Frost[2]

I met JJ again at a mathematics conference sponsored by Indiana University in Bloomington, Indiana. Just like the first time, he was sitting alone at a table in a corner of the hotel bar, gazing at me across the room with dark, inscrutable eyes.

As soon as I sat down, he pushed a piece of paper in my direction. "Here," he said laconically. "For your students."

I glanced at the paper and smiled. The text read something like this.

Question 1 (requires basic arithmetic or algebra). *If 3 farmers plow 3 hectares in 3 days, how many farmers are needed to plow 9 hectares in 9 days? Assume that each farmer works at the same rate and for the same length of time each day.*

Question 2 (requires common sense). *The tortoise intends to challenge the hare to a race, thinking that, if he is allowed a head start, he can beat the hare. This is based on the following argument.*

Let L_1 *be the starting point of the tortoise (ahead of the hare's). The signal is given and the race begins. When the hare reaches* L_1, *the tortoise has advanced to a new point* L_2; *when the hare reaches* L_2, *the tortoise has advanced to a new point* L_3, *and so on. Therefore, the tortoise believes that he will always be ahead of the hare, and so he will win the race.*

Is the tortoise correct in his thinking?

"Interesting," I said. "Thanks." I folded the paper and put it in my pocket, waiting for JJ to broach, as usual, a subject of his choice. He needed no special invitation.

"All your college freshmen," he said abruptly, "used to be taught proper calculus."

---

[1] See footnote 1 on p. 9.
[2] American poet, 1874–1963.

"They still are."

"Some are, and some aren't. In the 1980s, my colleague Gan attended a regional meeting of the Mathematical Association of America, where a couple of people spoke of the need for change in the way calculus was taught. The rationale, as explained to the audience, sounded not unconvincing: it was pointless for students to learn basic mathematical concepts like the derivative and integral as mere abstractions; they should do it in a way that relates such concepts directly to their applications; they should develop the skill to compute them from numerical data collected in the real world through experiment; they should perform the computation with an electronic calculator and not by hand from an analytic formula; and so on. There seemed to be great enthusiasm for the idea at the time. As we very well know, nothing pumps the academic adrenaline like the prospect of a revolution."

"Yes, I think it was a revolutionary idea," I said.

"Except that its champions did a very unBRITE thing: they abandoned logic by failing to take into account the reaction of the intended customers. We have followed events with interest from the moment the reformed calculus project got off the ground and its high priests and their confederates started implementing it in various parts of the country. Watching developments from the sidelines, the entire Jovian contingent was deeply apprehensive because, in the dim and distant past, our own civilization went through the same phase. Could it evolve differently in yours? we wondered. Could you succeed where we failed? Or would you have to fight hard, just as we did, to regain what you had before this expensive mistake was made?"

"What are you talking about?"

"I'm talking about mathematical sanity. I'm talking about freeing your mathematics from a terrible yoke that has been squeezing out its breath and stifling its progress. I'm talking about regaining the lust for, and the capability of, going to the stars."

"Aren't you a bit too melodramatic?"

JJ turned slightly sideways and crossed his legs. "This is no exaggeration. We've been so concerned that on one occasion Gan threw caution to the wind and, risking an accusation from the Ganymedean authorities of meddling in another planet's affairs, fixed it for himself to appear on a TV discussion panel on higher education, where he told the viewers why reformed calculus was not the most suitable way to introduce the students to differentiation and integration."

"He probably wasted his time. The great majority of the public hasn't even heard of calculus."

"Well, at least he tried. Here is the central part of Gan's exchange with the program moderator, Mod."

GAN: The teaching of calculus has taken a wrong turn in universities and colleges.

MOD: Excuse my ignorance, but my math knowledge stops at college algebra. Can you, in simple words, make me understand what calculus is about?

GAN: It's not easy, so I'll give you the bare bones. Consider a line. You may look upon it either as a set of points, which is the *discrete* view, or as an uninterrupted shape, which is the *continuous* view. Problems arising from the former are normally dealt with by algebra; those connected with the latter need the application of calculus. Differentiation, for example, studies changes in certain quantities, called functions, when other quantities, called variables, alter their values continuously.

MOD: I think I get the idea. Now what is *reformed* calculus? How does it differ from the unreformed version, and why do you think it is less successful as a teaching instrument?

GAN: Traditional calculus is taught by the instructor, who proves some of the mathematical statements in the classroom and shows examples of their application. The students are supposed to learn the material through individual, pen-and-paper practice sessions. The reasonably good ones succeed and go forth; the others keep trying again and again until they pass or give up.

Reformed calculus relies a lot less on teaching by the instructor, who is, however, on hand to answer questions. The students are subjected to the principle of constructivist learning: they work in groups, trying to solve problems and to arrive at definitions and methodology through collective exploration. They do this by using calculators or computers for all formal numerical or analytic manipulation, not bothering much, if at all, with rigorous proof or the correctness of the final result.

MOD: And what's wrong with that if, in the end, they get there?

GAN: What's wrong? The system is illogical! The students are expected to learn mathematics while they are struggling to apply it. Have you seen anyone able to play ice hockey before he can skate?

MOD: But trying to solve practical problems is a good thing. Isn't it?

GAN: The so-called real-world problems that the students are told to solve do not help them understand the mathematical principles lying at the basis of their solution. The step that takes them from concrete to abstract, essential in mathematics, is being suppressed. This might not be too bad for those with an inadequate mathematical background, but places at a disadvantage, and even risk, all those who progress to post-calculus courses.

The advocates of reformed calculus use mumbo-jumbo sound bites like 'higher-order thinking' and 'conceptual understanding', but at the end of the day, their system is nothing more than a kind of educational communism, where the strong do the work and the uninterested ride on the others' coattails and share in the results. And, just as communism as a political system doesn't seem to work satisfactorily, neither does this.

MOD: Dr. Gan, I think it would help us and our viewers a lot if you could illustrate the difference between traditional and reformed calculus by means of some simple analogy. Can you do that?

GAN: I'll try. Suppose that you take a person to a vast, completely dark room filled with all manner of boxes, and tell him that he must make his way to a door cut in the opposite wall. Totally disoriented, he starts fumbling around, shifting boxes this way and that, hitting his knees and elbows on hard corners, and moving randomly without a sense of direction until, if he perseveres enough and is lucky enough, he finally emerges at the designated target. You then switch on the light and show him his path across the room: a long, multisegment line with many unnecessary twists and turns, laboriously carved through a jumble of scattered packages. This is the reformed calculus method. In classical calculus, you first turn on the light and *then* ask the person to get to the door on the other side of the room. See the difference?

"After he got appointed at his university," JJ continued, "out of sheer curiosity, Gan sat in a few classes of reformed calculus. What he saw filled him with foreboding. The professors worked the projector-linked computer and covered the screen with arrays of numbers, pointing and commenting. The students, clumped in threes and fours, bewildered by all that, tried to do the same with their calculators. Looking at their faces, though, Gan guessed that they were not having a great deal of success or enjoying it.

"The following year, the survivors enrolled in Gan's differential equations course. The first time they met, Gan stated that calculus was a prerequisite for the course and that, since they were there, they must be fully conversant with differentiation and integration. He then proceeded to talk about the interpretation of the derivative as a rate of change. The students' reaction confirmed his worst fears. In the days before the reform, more than half of them would have accepted his explanation matter-of-factly, he said, as something they had already heard, even if they could not remember every detail. Not now. This time, only a handful seemed to understand what was going on; the rest looked void of comprehension. Gan's VOTSITed dialogue with Stu on that occasion is quite revealing."

GAN: Let's begin with the definition of the derivative of a function.

STU (looks at what Gan has written on the board): I'm sorry, sir, but I don't think I've seen this before.

GAN: What do you mean? What definition were you given in your calculus course?

STU: When we learned about the derivative, I remember that there were lots of numbers, and that we messed around with them on our calculators, nor sure where we were going. We even asked each other if we were in a real mathematics course. We couldn't see the mathematics.

GAN: But you must've been given the analytic formula as well.

STU: If we were, it probably got drowned in that sea of numbers. Personally, when I learn something new in math, I like to start with a general formula, backed up, if necessary, by a supporting sketch. This gives me an intuitive handle on the concept. I can visualize it. And once I have that, I can play safely with numbers and approximations in the lab, because I know where I'm going and I know what method to use. I thought mathematics was about clear, abstract thinking, not about getting lost in endless numerical confusion.

GAN (sees a majority of the students vigorously nodding agreement): I'm afraid you have suffered 'ordeal by reformed calculus'.

STU: Well, as far as I'm concerned, whoever came up with this reform should be bought a one-way passage to Epsilon Eridani.[3] Maybe the folks in that system will find their invention useful.

GAN (almost says "No, they won't" but catches himself in time): Your unhappiness has been noted.

STU (a hint of reproach in his voice): Why was the reform necessary? Older scientists and engineers learned their calculus unreformed, and they had no problem building a great country with it.

"Gan couldn't blame the students," JJ said, "and he couldn't blame the teachers, who were excellent. But he could, and did, blame the colleagues responsible for jumping on the bandwagon and forcing everyone to worship the false GAWD."

"Did he remonstrate with them?"

"Many times—to no avail. They remained unmoved. They said that the solution was for him to teach *reformed* differential equations as well. It took a lot of self-control to stop Gan from throttling one or two people. 'For as long as I work here,' he told them, 'the use of electronic computational devices by students in my classroom will be reason for failure. Calculators are learning-inhibitors. I will not make a mockery out of mathematics. I will not tarnish its beauty and majesty.'"

"Don't his students complain about the ban on calculators?" I wondered.

"Why should they? The data in the questions on Gan's tests are small integers, easily managed by hand."

"Real-life data aren't like that."

"All science and engineering students learn programming in one language or another. When they go on to face real life they'll write codes to instruct the machine to do the boring computations for them. They *will* be able to do that because Gan and others of like mind have, with simple examples, taught them how to find the solution of the problem, which a computer, by itself, will never do."

"Unfortunately, schoolchildren are allowed and encouraged to use calculators for every sum, however trivial," I lamented.

JJ nodded. "The teaching of mathematics has been pushed through radical reform from top to bottom, and it hasn't improved. Quite the opposite. Some students even use the word 'reformed' as a term of derision. Listen to the conversation Gan had with one of his sophomores, Sop, after the fall semester's finals."

SOP: I'm told that last year we learned *reformed* calculus. If this department dishes out reformed mathematics, why doesn't it also embrace reformed logic?

GAN: What do you mean? What kind of logic would that be?

SOP: Well, let's take your class of differential equations. A student who scores an average of 90 gets an A, right?

---

[3] A star in the constellation Eridanus.

GAN: Of course.

SOP: But what if he scores 89.5? Such a minute difference could easily be explained by a slight grading unevenness, due, for example, to grader fatigue caused by a very stressful day at the office. Would you not, in that case, consider giving the student an A? In the interest of justice and fairness?

GAN: I probably would.

SOP: Okay, so we have established that 89.5 is an acceptable A standard. Suppose now that the student's average is 89. Arguing as before, would you not say that the student still deserves an A?

GAN (seeing what Sop is getting at): What are you getting at?

SOP: Using this reformed logic, sir, we can conclude that someone who scores 75 is also worthy of an A.

GAN (amused): How much did you score in my class?

SOP: 75.

GAN: You realize that this reasoning system works the other way as well, don't you? We know that 59 is an F. Then, according to the Principle of Truth by Close Proximity that you just formulated, 59.5 would also be an F. And so would 60. Continuing the chain, we would then conclude that someone with 75 has failed. Now what would you have me do: leave you with the C you earned, or use your reformed logic to determine your final grade? Before you answer, remember that, being the professor, it's up to me to decide in which direction that logic will be applied.

"The intention behind reformed calculus was good, though," I said.

"So was, in its time, the intention behind geocentrism. But that idea turned out to be no more helpful to a true understanding of the relative motion of celestial bodies than this reformed nonsense is to assimilating the fundamental concepts of differentiation and integration. It seems, however, that things are on the mend. A few years ago, Gan's university reintroduced an alternative stream of traditional calculus. Gradually, as more and more students began choosing the classical version over the reformed one, the latter lost ground to the extent that next year, Gan tells me, its teaching will be discontinued. I'm also pleased to note that many colleges appear to be going the same way. The hydra is having its heads cut off, one by one. The Seventh Mathematical Commandment is adhered to by increasing numbers in the profession. My friend, there is hope for your civilization yet." JJ leaned back in his chair. "Now," he said, "it's time for your stories."

I finished the last of the coffee and pushed the empty cup aside. "OK, here we go. First one coming up."

A man dies and, after going through the customary check at the Pearly Gates, is sent down to Hell. Annoyed that he hadn't made the cut, he appears before B.L.Z. Bub, who tries to console him:

"Don't be disappointed. You'll see that being here is not too bad. Every day we do different things, all of them most enjoyable. You like smoking?"

"Yes, I love smoking."

"See? Every Monday we do nothing but smoke. And you don't have to worry about your lungs killing you because you're already dead. You like drinking?"

"Absolutely."

"This is what we do every Tuesday. Whether your liver takes it or not, why should you care? You're dead, remember? So drink to your heart's content. You like food?"

"Oh, man, do I!"

"Excellent. On Wednesdays we eat like crazy. You'll get all the food you want, in unlimited quantities. Obesity and heart attacks do not occur in the land of the dead. You like gambling?"

"Sure. Don't tell me—every Thursday?".

"Without fail. Money grows on trees around here, and there's no limit on stakes."

The man starts feeling more reassured and thinks that Hell is a pretty comfortable place.

"You like reformed calculus?"

"I can't stomach it. That thing always gave me terrible headaches and made me queasy."

B.L.Z. Bub looks at the man with pitying eyes. "Then you'll hate Fridays," he says. "Particularly those with tests."

"You know how to make a person happy!" JJ exclaimed. "Even a nonhuman!"

"Yeah, I thought you'd like it. And now," I said, "for a little change of nuance, a story that demonstrates the versatility of a calculus test when used as a connector between instructor and student."

Four undergraduates in the same calculus class, good friends and living in the same neighborhood, realize that they are not sufficiently prepared for their midterm exam and work out a plan to improve their chances of getting a decent grade. Thus, on exam day, they skip class and after the exam has finished go to see their professor.

"Sir," one of them explains, "we are very sorry we missed the test, but there's a good reason. We all travel to the university in the same car, and this morning, just as we were turning from the driveway into the street, one of the tires blew out. We replaced the wheel with the spare as fast as we could, but did not manage to arrive here on time. Would it be possible for you to give us a makeup?"

"Yes," the professor says. "You have a valid excuse. Come to my office next Monday at 10 am and I'll have a make-up paper ready for you."

The students are very happy: not only have they been allowed extra time to prepare, but, by looking at the paper given on that day, they also got a pretty good idea which part of the material the professor favored.

At 10 am the following Monday, the four students are at the professor's office.

The professor gives them the exam paper and answer sheets, indicates four doors along the corridor, and says, "These offices are empty." Then, addressing one student at a time, he points, "You go into this one, you into that one, you into that, and you into that. I'll come and collect the answers in one hour."

The four students enter their assigned offices, sit down at the desks inside, and read the paper:

CALCULUS 102

1. Compute
$$\int_0^1 2x\,dx.$$
(5 points)

2. Sketch the car in which you attempted to drive to the college last week and mark the tire that blew up.

(95 points)

"One of the biggest mistakes students make is to believe that they can outsmart their teachers." JJ put his hands on the table. "Thanks for the stories," he said. "I've really enjoyed our little chat. If all goes well, we'll talk some more next time." Then, unceremoniously, he sprang to his feet and promptly walked out of the bar.

Notes After the Meeting

Here are the answers to JJ's questions together with some brief personal comments.

Solution to Question 1 (arithmetic). There are three quantities involved in this problem: the number of workers, the worked area, and the working time. Under the given assumptions, it is clear that:

(i) if the working time remains constant, the worked area is proportional to the number of workers (that is, the more workers, the larger the area worked in the same interval of time);

(ii) if the worked area remains constant, the working time is inversely proportional to the number of workers (that is, the more workers, the less time it takes to work the same area).

The following reasoning chain is now easily constructed:

$$\begin{aligned}
&\quad\; 3 \text{ hectares are plowed in } 3 \text{ days by } 3 \text{ farmers} \\
&\Rightarrow\; 1 \text{ hectare is plowed in } 3 \text{ days by } 3 \div 3 = 1 \text{ farmer} \\
&\Rightarrow\; 1 \text{ hectare is plowed in } 1 \text{ day by } 1 \times 3 = 3 \text{ farmers} \\
&\Rightarrow\; 9 \text{ hectares are plowed in } 1 \text{ day by } 3 \times 9 = 27 \text{ farmers} \\
&\Rightarrow\; 9 \text{ hectares are plowed in } 9 \text{ days by } 27 \div 9 = 3 \text{ farmers}.
\end{aligned}$$

Hence, the answer is 3.

Solution to Question 1 (algebraic). We denote by N the number of workers, by A the worked area, and by T the working time. Remarks (i) and (ii) above are modeled by the formula

$$N = k\frac{A}{T}, \qquad (9.1)$$

where k is a proportionality constant. With hectares and days as units for area and time, the full given data set

$$N = 3, \quad A = 3, \quad T = 3$$

replaced in (9.1) yields

$$k = \frac{NT}{A} = \frac{3 \times 3}{3} = 3,$$

so the exact form of (9.1) is

$$N = 3\frac{T}{A}. \qquad (9.2)$$

Hence, for $A = 9$ and $T = 9$, from (9.2) we obtain

$$N = 3 \times \frac{9}{9} = 3.$$

When they are asked this question, some people do not bother to reason it through and, 'persuaded' by the symmetry of the data, give the answer 9, which is incorrect.

Solution to Question 2. The tortoise is wrong because he is considering motion as being described only by distance. When time is also thrown in, the picture changes completely. The distances from L_1 to L_2, from L_2 to L_3, and so on, are traveled in shorter and shorter intervals of time, and this discretized time process converges to an instant when the tortoise is caught up. Time, being a continuous unbounded variable, flows beyond that instant, quickly putting the hare out of sight.

It is hoped that the tortoise gave up the idea of the race as a bad job.

The original of this problem pits Achilles[4] against the tortoise, and it is one of the paradoxes (aporias) proposed by Zeno of Elea.[5]

A Word to the Wise

Operations with Fractions

All denominators in this section are assumed to be nonzero.

---

[4] A Greek mythology hero, son of king Pelleas of Thessaly and the nymph Thetis; reputedly a very fast runner.
[5] Pre-Socratic Greek philosopher, c490–430 BC.

☺ DO

Here is a list of basic operations:

$$\frac{a}{b} + \frac{c}{d} = \frac{ad}{bd} + \frac{bc}{bd} = \frac{ad+bc}{bd},$$

$$\frac{a}{b} \times \frac{c}{d} = \frac{ac}{bd},$$

$$\frac{a}{b} \div \frac{c}{d} = \frac{a}{b} \times \frac{d}{c} = \frac{ad}{bc},$$

$$\frac{a+b}{c} = \frac{a}{c} + \frac{b}{c},$$

$$\frac{ac}{bc} = \frac{a}{b}.$$

Two other important properties of fractions are

$$\frac{a}{b} = \frac{c}{d} \quad \Leftrightarrow \quad ad = bc,$$

$$E = \frac{a}{b} = \frac{c}{d} \quad \Rightarrow \quad E = \frac{a+c}{b+d}.$$

'Legal' simplification of fractions is a reflex for mathematicians. When you see someone working with a ratio like 6/8 instead of the simpler, equivalent version 3/4, you will know immediately that he is not a true mathematician.

☠ DON'T

The first two 'equalities' below are wrong, as they are in direct violation of JJ's Second and Third Mathematical Commandments:

☠ $\quad \dfrac{a+b}{a+c} = \dfrac{\not{a}+b}{\not{a}+c} = \dfrac{b}{c},$

☠ $\quad \dfrac{a}{b+c} = \dfrac{a}{b} + \dfrac{a}{c},$

☠ $\quad \dfrac{a}{b} + \dfrac{c}{d} = \dfrac{a+c}{b+d}.$

Although I have not come across the next pair myself, I have been assured by others that they are genuine documented errors:

☠ $\quad \dfrac{\sin x}{\sin y} = \dfrac{\sin\!\!\!/\, x}{\sin\!\!\!/\, y} = \dfrac{x}{y},$

☠ $\quad \dfrac{\sin x}{x} = \dfrac{\sin \not{x}}{\not{x}} = \sin.$

The last 'equality' especially is indeed an unpardonable sin!

SCAM 10

Political Correctness

> *Political correctness is the natural continuum from the party line. What we are seeing once again is a self-appointed group of vigilantes imposing their views on others. It is a heritage of communism, but they don't seem to see this.*
> Doris Lessing[1]

> *Don't you see that the whole aim of Newspeak is to narrow the range of thought? In the end we shall make thought-crime literally impossible, because there will be no words in which to express it.*
> George Orwell[2]

I met JJ again at a mathematics conference sponsored by The Johns Hopkins University in Baltimore, Maryland. Just like the first time, he was sitting alone at a table in a corner of the hotel bar, gazing at me across the room with dark, inscrutable eyes.

As soon as I sat down, he pushed a piece of paper in my direction. "Here," he said laconically. "For your students."

I glanced at the paper and smiled. The text read something like this.

Question 1 (requires elementary logic). Four people, P, Q, R, and S, must walk across an unlit bridge at night under the following conditions:

(i) they are initially on the same side of the bridge;

(ii) they have only one flashlight between them;

(iii) at most two people can cross together at any one time;

(iv) any crossing party must have the flashlight with them;

(v) the flashlight must be walked across the bridge (it cannot be thrown from one side to the other);

(vi) each of the four people walk at a different pace: P can complete the cross in 1 minute, Q in 2 minutes, R in 5 minutes, and S in 10 minutes;

(vii) any pair must walk together at the slower person's pace.

---

[1] British writer, b1919.
[2] The pen name of British author and journalist Eric Arthur Blair, 1903–1950.

What is the shortest time in which the entire group of four can transfer from one side of the bridge to the other?

Question 2 (requires logic). If arithmetic is a consistent system (that is, if it does not generate any contradictory statements), can it contain true theorems that cannot be proved to be true?

"Interesting," I said. "Thanks." I folded the paper and put it in my pocket. This time I didn't wait for JJ to broach a subject of his choice. "Want a cup of coffee? A beer?"

"I don't drink coffee; or alcohol."

"So what does a Jovian–American drink?"

JJ gave me a strange look, a marked departure from his serene, unperturbed countenance. "Don't get cute with me."

"My question upset you!" I stated the obvious.

"I'm not an American. I'm not even an Earthman. But if I were, I would object most vehemently to being called a *Jovian*-American."

"Why? You were born on one of Jupiter's moons, weren't you? Calling you a Jovian–American would reflect the truth."

"Calling me a Jovian–American would be discriminatory—it would imply that I wasn't really a fully fledged American, that there was something less than American about me. In my view, this kind of politically correct labeling is a very unBRITE practice, smacking of illogic."

"Yet many people adopt it because they take exception to the explicit mention of their race, religion, height, or weight. How would you describe a person without giving offense?"

"I admit that certain words can hurt. Since they are used to express thoughts and feelings, they have come to be closely associated with those thoughts and feelings, which are not always honorable. Isolated letters and numbers, on the other hand, are neutral and impersonal. So why not introduce a simple code that ascribes letters or numbers to gender, place of birth, color, age, height, weight, and religion, say, alphanumerically? For example:

gender:	1 female, 2 male
place of birth:	1 Africa, 2 America (Central or South), 3 America (North), 4 Antarctica, 5 Asia, 6 Australia, 7 Europe, 8 Other World
color:	1 black, 2 blue, 3 brown, 4 green, 5 orange, 6 pink, 7 red, 8 white, 9 yellow
age:	a 0–5, b 6–10, c 11–15, d 16–20, ...
height (cm):	a 0–50, b 51–100, c 101–120, d 121–130, ...
weight (kg):	a 0–50, b 51–55, c 56–60, d 61–65, ...
religion:	1 agnostic, 2 atheist, 3 Buddhist, 4 Christian, 5 Jewish, 6 Muslim, 7 taoist, 8 other.

"Then you establish an order for quoting the most important physical characteristics of a human—something like (i) gender, (ii) place of birth, (iii) color of skin, (iv) color of hair, (v) color of eyes, (vi) age, (vii) height, (viii) weight, (ix) religion—allocate 0 to mean 'unsure', 'undetermined', or 'withheld', and refer to all persons by their corresponding code. Precise, clean, unbiased, logical, and pleasing all-around."

I couldn't resist, so I asked, "What would yours be?"

Without hesitation, JJ said, "28431vfb0."

"Let me see: you are a male extraterrestrial with green skin, brown hair, and black eyes—did you say 'v' for age? What, you are 105 years old??"

"107 in your units, to be precise. Also, I'm 149 cm tall and weigh 52 kg."

He looked not a day older than 45, but it's really hard to tell with ET, isn't it? "You gave 0 for religion," I said. "How come?"

"I don't have a religion as such."

"Anyway," I insisted, "what is it you believe in?"

"Mathematics."

I should have seen it coming. "The sort of code you propose is unwieldy," I said. "It wouldn't catch on here."

"Why not? People would get used to it very easily, the way they do with telephone numbers. Of course, you may shorten the string. If you were to note only the gender, place of birth, color of skin, and height, then my code would be 294f. Much neater than a little green man from outer space."

Curiosity got the better of me. "Have you been involved in any political correctness incident since you assumed human shape?"

"Not myself personally. But Gan had a couple of weird experiences, and both were about gender. This is what happened between him and a female student, Fem, when he wanted to give his calculus class a simple example of a max-min problem. Gan always records his lectures with the VOTSIT.

GAN: Consider a man on an oil rig 4 km from a straight shore. He can row at 3 km per hour and run at 12 km per hour—

FEM: Why don't you pick a gender-free example? Why a man? You think a woman is incapable of working on an oil rig?

GAN (nonplussed): I'm sure a woman would be perfectly fine out there, but I didn't want to make her row a long distance and then run along the shore an even longer one. I'd rather have a man go through that physical ordeal.

FEM (not satisfied): Everything a man can do a woman can do, too. And more.

GAN (conciliatory): Very well, let's take it again from the top. A *woman* on an oil rig—

FEM (annoyed): You've changed the worker's gender only to appease me, not because you really agree with my point.

GAN (wants to end this unnecessary sparring): Here is a neat compromise. Let's not have a woman, or a man, or even a dog on that rig—we don't want to be accused of subjecting an animal to unnecessary cruelty; let's have a sentient entity. And if this still displeases you, then we'll ask your colleagues what they prefer: to hear me explain the solution to a problem that might be on the test next week, or to witness the discussion of an issue which will definitely *not* be on the test.

"There is nothing like the concerted accusatory looks of 50 peers to make one stop badgering the professor with a sterile argument," JJ said.

"And the second incident?"

"It occurred last year, at the end of the commencement ceremony in Gan's institution. The faculty were filing out of the hall two by two, all robed and silent. The exit from the hall was through a narrow gangway that would take them over to the other side of the building. Gan's pair in the slow procession was Wom, a 17893hfi0 according to my proposed code, whom he didn't know."

"Wait a minute," I interjected. "If her height and weight are quantified by 'f' and 'i', then—"

"Exactly. Wom was what in PC terminology you might call a vertically challenged metabolic underachiever. She and Gan had a brief exchange as they arrived at the door."

GAN (motions with his hand): After you, please.

WOM (pugnaciously): Why are you doing this? Because I'm a woman?

GAN (taken aback): Oh, no. Not at all. It's—

WOM (incensed): You think women are some kind of weaklings? You think we need special privileges and concessions?

GAN: The thought never crossed my mind. Would you rather I stepped out first?

WOM: Assertion of male dominance is equally unacceptable.

GAN: Then we're stuck. Let's just remain where we are until one of us dies, shall we?

WOM (angry and disdainful): Your sarcastic attitude is contemptible. So typical of a man! The solution is obvious: we will go through the door *simultaneously*. Why didn't *you* suggest it? Too practical for your mind?

GAN (overlooks Wom's sarcasm): Contrary to what you may believe, that was also my first idea. But I rejected it on logistic grounds. Your diameter-to-height ratio, which ordinarily would matter not a bit, in this particular instance becomes hugely significant. You and I will not fit through this door together, and, since I'm in no hurry, I'll wait until you decide how you want to proceed.

"What did Wom do?" I asked.

"She said nothing and walked out ahead of Gan."

"It seems that Gan came across a truly radical feminist. Talking about which, how do you, Jovians, view feminism?"

"We think it's a natural response to a very real problem. In some parts of the world women are treated like chattel, which is shocking and inexcusable. But even in the developed countries sex discrimination has a relatively high incidence. Not long ago, a female Gan documenting such cases was interviewed by a company that had advertised for a mechanical engineer. Watch the VOTSIT recording of her conversation with the staff partner, Par."

PAR: What is your highest education achievement?
GAN: A Ph.D. from CalTech.
PAR (smiling): You must've learned a lot of theory about machines. With all that wonderful knowledge, can you repair a car?
GAN: No.
PAR (smiling more widely): I thought not.
GAN: But I can design you one.
PAR (smiling less widely): Thank you for coming. Unfortunately, you are not the right person for us.
GAN: Then why did you bother to interview me? After all, you had read about my credentials on the application form.
PAR (still smiling): I just wanted to see what a woman engineer looked like.
GAN: And are you satisfied?
PAR: Yes.
GAN: Before I go, can you do me a favor? Would you simplify this for me? (Writes $\sqrt{x^2 + y^2}$ on a piece of paper.)
PAR: Why? Didn't learn enough math at CalTech? (Picks up the pen and, confidently, writes $= x + y$.) There.
GAN: Now I'm also fully satisfied myself.
PAR (a little confused): What do you mean?
GAN: You've just confirmed what I suspected all along. You never heard of the Fifth Mathematical Commandment. Your MICQ is no higher than 24. To be perfectly honest, I was quite certain that you wouldn't hire me, but I wanted to see what a male chauvinist looked like.

"Brilliant!" I approved. "Gan had the right answer."

"Women's cause is a vital issue for humanity, but it's not clear if feminism, in its present form, is helping it effectively. Women should be encouraged and supported to do what *they* want to do, not what others tell them to do. In many parts of the world, a strong, talented, determined, well-motivated woman has every chance to build a successful career and beat men at their own game. Margaret Thatcher[3] quipped that if you want something said, ask a man; if you want something done, ask a woman. Herself is an excellent example. She didn't need any feminist prop to raise two children, get a degree in chemistry, qualify as a lawyer, and be elected to the highest office in the U.K.—three consecutive times, no less. She took a country left by her *male* predecessors in a sorry state and turned it around, becoming one of the most influential political leaders of the last century. A fact that cannot be denied, whether you liked the color of her politics or not."

"The incidents you described earlier are not very common. In general, people behave sensibly and avoid extremism."

"Some do, and some don't. Anyway, I think that political correctness is a sham: it replaces one discriminatory system with another and impedes open and efficient communication between the members of your society."

---

[3] Former British prime minister, b1925.

"Do you see political correctness affecting mathematics? That is, beyond the demand for gender-free examples?" I asked.

"There are 'critical theorists' who assert that mathematics should be used to promote social justice. They proclaim that the mathematics taught in our schools and colleges reflects the values of western civilization and, therefore, it serves the purposes of conquerors and oppressors. In other parts of the world, they say, mathematics should be taught in accordance with the history and traditions of those other cultures. They call this 'ethnomathematics', but I can think of more appropriate names for it. If things are allowed to get out of hand, I wouldn't be surprised one day to hear that primes should be referred to as 'divisor-challenged', even numbers as 'differently odd', infinity as 'boundedness impaired', wrong reasoning as 'alternative logic', exam failure as 'deficiency achievement', and so on." JJ leaned back in his chair. "Now," he said, "it's time for your stories."

I finished the last of the coffee and pushed the empty cup aside. "OK, here we go. Fem and Wom would be very displeased with the first one. Ironically, I heard it from a woman."

A student, annoyed that his girlfriends are dumping him regularly, tries to rationalize the phenomenon and applies mathematics to find its cause (which, he suspects, has nothing to do with him). Thus, based on direct observation and experience, he starts from the assumption that girls require both time and money be spent on them, which he expresses by means of the formula

$$\text{girls} = \text{time} \times \text{money}. \tag{10.7}$$

But, as is well known, time *is* money:

$$\text{time} = \text{money}.$$

Replacing this in (10.7), he obtains

$$\text{girls} = \text{money} \times \text{money} = (\text{money})^2. \tag{10.8}$$

Next, another piece of conventional wisdom has it that money is the root of all evil, so he writes

$$\text{money} = \sqrt{\text{evil}}.$$

He substitutes this in (10.8) and arrives at the equality

$$\text{girls} = \left(\sqrt{\text{evil}}\right)^2 = \text{evil}.$$

Based on this irrefutable proof that girls are evil, he dismisses any lingering thoughts of personal inadequacy and, with renewed determination, starts devoting his spare time again to the pursuit of the fairer sex.

"Trying to find an exception to the rule, even when mathematics shows this to be impossible, is an exciting challenge for the young," JJ said.

"Numerically speaking, however, in some people's minds the root of all evil is approximately 25.806976," I quipped.

JJ's eyes twinkled briefly. "Ah, forget that silly old superstition. I guess you also have something in the opposite direction? Something to please Fem and Wom?"

"You guess well. Here it is."

A beautiful young princess sits on the lawn of her castle, reading with great interest the prospectus received from the Department of Environmental Engineering at the local university. Suddenly, a frog jumps in her lap and says, "Sweetie-pie, I am a handsome prince who had the misfortune to be transformed into this repulsive creature by a wicked witch. If you kiss me, I will immediately become my former self and marry you. Then I will bring my mother and we will live together in your castle, where you will cook for me, wash my clothes, bear my children, and feel eternally grateful for the opportunity to serve me so."

An hour later the princess is dining on grilled frog legs sprinkled with a savory sauce made of organic herbs and polyunsaturated oil. As she washes down the last mouthful with a few sips of alcohol-free wine, she whispers to herself, happy and satisfied, "You wanted me to serve you, my gentle prince? Well, there you are: all served up."

"A modern and very liberated approach to an old issue." JJ put his hands on the table. "Thanks for the stories," he said. "I've really enjoyed our little chat. If all goes well, we'll talk some more next time." Then, unceremoniously, he sprang to his feet and promptly walked out of the bar.

Notes After the Meeting

Here are the answers to JJ's questions together with some brief personal comments.

Solution to Question 1. This is based on two simple observations:

(a) There will be a crossing completed in 10 minutes, because of S, so it makes perfect sense to group the two slowest persons (R and S) into one crossing party.

(b) The flashlight will have to be carried back and forth by one of the fastest persons, that is, P or Q.

Then the reasoning becomes straightforward:

1. P and Q make the first crossing; duration: 2 minutes.

2. P crosses back alone with the flashlight; duration: 1 minute.

3. R and S pick up the flashlight from P, leave P behind, and cross together to join Q; duration: 10 minutes.

4. *Q* crosses back alone with the flashlight to join *P*; duration: 2 minutes.

5. *P* and *Q* cross together to join the other two on the far side; duration: 2 minutes.

The minimum crossing time for the group is therefore $2 + 1 + 10 + 2 + 2 = 17$ minutes.

The usual solution given by various audiences is 19 minutes, which is based on the premise that P, the fastest in the group, should grab hold of the flashlight and take the other three across one at a time.

There is a second version of the solution which yields the correct answer 17; the readers should try and find it themselves.

Sometimes, the odd customer comes along who, in spite of strenuous assurances that this is a genuine mathematical problem, looks for a trick and considers all manner of bizarre possibilities: a hidden boat, or a dry riverbed under the bridge that can be traversed on foot by all four together, or two of the persons being carried piggy-back by the other two (which would technically mean that only two were *walking*)—to name but a few. Such oddball ideas are a waste of time.

I once had a bunch of students with a fine sense of humor in one of my courses, with whom I discussed a largely neglected aspect of the problem: the political correctness of its formulation. Our debate highlighted a number of issues and questions that, sooner or later, will probably be fully addressed by modern educators.

(i) The text discriminates against less sprightly people like R and S. This is called *slowism*.

(ii) If P can walk across the bridge in 1 minute while S needs 10 minutes for the same distance, it is reasonable to assume that S is probably differently abled and should not be forced to walk at all. A wheelchair or a walker should be provided for S, and the text should acknowledge that. The case of R should also be examined.

(iii) The local authority should be petitioned to provide adequate lighting along that footbridge, to make it accessible at night also to pedestrians who cannot afford to buy a flashlight. Alternatively, all such people living within a 10-mile radius of the bridge should be supplied with free flashlights.

(iv) Since only at most two persons can cross the bridge at the same time, it has to be concluded that the bridge is kind of rickety and therefore unsafe for pedestrian traffic. It is likely that one day the bridge will collapse, harming not only people but fish, waterfowl, mosquitos, and other endangered species living under it.

(v) Why are these four persons being subjected to the hazard of walking at night across an unlit, poorly maintained bridge? How important is the task awaiting them on the other side that they must cross in the shortest possible time? What nationality are they? If they are aliens, are they legal? Will these people be paid for the task they will perform? Has the appropriate union been informed? Does the IRS know about it?

We ended our debate by reformulating the problem in a way that, we thought, might take account of these issues and therefore satisfy most of the people most of the time. The modified version sounded completely crazy.

One final note: literal enumerations invariably start with $A, B, C, \ldots$, so, in order to avoid accusations of discrimination against the letters closer to the end of the alphabet, I have named the persons in the problem P, Q, R, and S. This, as is easily seen, also avoids the pitfall of making the text gender-specific.

Solution to Question 2. This question was raised in 1931 by Kurt Gödel,[4] who answered it in the affirmative with an elegant argument that involves the following assertion.

Theorem. *This theorem cannot be proved.*

Proof. *Suppose that the theorem is false. Then the opposite of what it says must be true, that is, the theorem can be proved. If it can be proved, then, according to the theorem itself, it cannot be proved. We have obtained two contradictory statements, which means that our system is inconsistent. However, since we have assumed the system to be consistent, we conclude that the assumption about the theorem being false is wrong; hence, the theorem is true, but, by its own assertion, it is unprovable.*

Gödel's original argument uses an ingenious technique that represents every mathematical equation and assertion by a unique positive integer.

It is unfortunate that this type of argument does not help us identify which true theorems cannot be proved. For a long time it was thought that Fermat's so-called last theorem[5] might be one of them, but in 1995 that suspicion was shown to be unfounded when a proof was constructed. Goldbach's conjecture[6] about primes is another possible candidate. Until, that is, someone, some day, settles that question as well, one way or the other. But even then we must not despair, for mathematics is a seemingly unbounded field and there is an awful lot of it that has not been thought of, let alone explored, yet.

Supplementary note: anyone wandering why JJ called 'a silly old superstition' my tongue-in-cheek statement about 25.806976 being the approximate root of all evil should pick up a calculator and use its squaring function to find the answer.

---

[4] Austrian-born American mathematician, 1906–1978.

[5] Pierre de Fermat: French lawyer and mathematician, 1601–1665. This theorem states that if n is an integer greater than 2, then there are no nonzero integers a, b, c such that $a^n + b^n = c^n$.

[6] The 'strong' Goldbach conjecture states that every even integer greater than 2 can be written as the sum of two (possibly repeated) primes. Christian Goldbach: Prussian mathematician, 1690–1764.

A Word to the Wise

Algebraic Identities

☺ **DO**

The following identities should be remembered by anyone studying mathematics beyond primary school:

$$(a+b)^2 = a^2 + 2ab + b^2,$$
$$(a-b)^2 = a^2 - 2ab + b^2,$$
$$a^2 - b^2 = (a+b)(a-b).$$

A few other useful ones are

$$(a+b)^3 = a^3 + 3a^2b + 3ab^2 + b^3,$$
$$(a-b)^3 = a^3 - 3a^2b + 3ab^2 - a^3,$$
$$a^3 + b^3 = (a+b)(a^2 - ab + b^2),$$
$$a^3 - b^3 = (a-b)(a^2 + ab + b^2),$$
$$(a+b+c)^2 = a^2 + b^2 + c^2 + 2ab + 2bc + 2ac.$$

☠ **DON'T**

Here are some completely nonsensical 'equalities':

☠ $(a+b)^2 = a^2 + b^2,$
☠ $(a-b)^2 = a^2 - b^2.$

Accepting either of them as valid would imply belief in the linear behavior of the square function and would constitute a direct violation of the Fifth Mathematical Commandment.

SCAM 11

TV Advertising

> *Advertising: the science of arresting the human intelligence long enough to get money from it.*
> Stephen Leacock[1]

> *It is pretty obvious that the debasement of the human mind caused by a constant flow of fraudulent advertising is no trivial thing. There is more than one way to conquer a country.*
> Raymond Chandler[2]

I met JJ again at a mathematics conference sponsored by Kansas State University in Manhattan, Kansas. Just like the first time, he was sitting alone at a table in a corner of the hotel bar, gazing at me across the room with dark, inscrutable eyes.

As soon as I sat down, he pushed a piece of paper in my direction. "Here," he said laconically. "For your students."

I glanced at the paper and smiled. The text read something like this.

Question 1 (requires basic arithmetic). *Father and son are aged 43 and 16. After how many years will the father be twice as old as his son?*

Question 2 (requires basic algebra). *Show that the number*
$$n^4 + 4^n$$
is not prime for any positive integer $n \geq 2$.

"Interesting," I said. "Thanks." I folded the paper and put it in my pocket, waiting for JJ to broach, as usual, a subject of his choice. He needed no special invitation.

"Why do humans advertise things on television?" he asked abruptly.

"Because they want to sell things, I guess."

"They can do it on the internet. Almost everybody in the developed world has access to a computer these days."

"Not the same impact," I said.

"I don't think it's just that. I think a lot of humans prefer to be fed information rather than search for it. TV advertisers spot the trend and take full advantage of it. But their creations are very uneven, ranging from inspired

---

[1] Canadian writer and economist, 1869–1944.
[2] American writer of detective fiction, 1888–1959.

to downright laughable. The BRITEs should boycott all products featured in lame, insipid, or offensive ads."

"Why?"

"There is a kind of simple logic at work here. Bad ads are made by bad advertisers; bad advertisers are hired by manufacturers/providers who can't recognize bad when they see it and are, therefore, bad themselves; and bad manufacturers cannot be trusted to make anything except bad products. Ergo,[3] by transitivity, one shouldn't buy an item pushed by a bad ad because chances are the item is bad. Remember what I said about untalented TV advertisers in our classification? They don't rate a MICQ higher than 24."

"What brought this on?" I asked.

"There was a commercial break on the TV just before you came in. A woman was calling the family doctor to ask what medicine she should give her daughter, who was ill. The little girl, sitting by her mother in their home's living room, looked anything but ill. In fact, she looked as if she had been promised a big dollop of chocolate ice cream if she would keep quiet for a minute, pretending to be in pain. The mother picked up two little plastic containers in one hand and, holding them in front of her eyes, said into the receiver, 'Doctor, I bought brands A and B from the pharmacy: which one should I give her?'"

"So?"

"*So?* Tell me, what normal, thinking person buys two different brands of medicine used for the same purpose *before* asking the doctor which one is better?"

Though calm and composed, JJ could make his arguments sound very forceful.

"Here is another one," he went on. "You listen to a man's voiceover presenting a new drug—let's call it phonycine—intended to shrink one's feet. After extolling its virtues, he then mentions that the drug may have side effects, which include loss of limb, bubonic plague, tsunami, and armageddon. Would you ever consider buying a product with possible side effects like that?"

"The manufacturers must cover themselves in case there is a mishap. They are trying to avoid damaging litigation," I said.

"If the drug has such horrendous side effects, it shouldn't be allowed on the market. If it's not safe, it ought to be be withdrawn and banned."

"The odd iffy drug may be the price we have to pay for pharmaceutical advancement."

"Phooey," JJ scoffed. "But the most amusing part of the ad is its ending, when the voiceover says, 'Talk to your doctor to see if phonycine is right for you.' Gan surveyed physicians and asked them how they would really like to respond if their patients did that. The answers of a typical doctor, Doc, to such a patient, Pat, left no room for doubt.

PAT: I heard about this new product, phonycine. Do you think it would be good for me?

---

[3] Therefore. (Latin)

DOC (calmly): How many years of medical training do you have?

PAT (confused): Well, none.

DOC: I spent four years in pre-med, then four in medical school, and then four as a resident before I was fully qualified to practice on my own. I read the medical and pharmacology journals regularly. I keep abreast of everything that's new. Why are you asking me if phonycine would be good for you? Don't you think that if I thought it were, I would tell you myself?

PAT: But the manufacturer's TV ad says—

DOC: If you are guided by the ad, then ask the manufacturer to treat you.

"Of course," JJ commented, "however irritated they may feel, physicians don't speak like that to their patients; they also enjoy the freebies they get from the drug manufacturers."

"Any other TV stuff that annoys you?"

"Plenty. Listen to what JOEs had to say in another survey, after Gan asked them for their immediate reaction when he described a number of specific ads."

GAN: A couple of characters are having an asinine exchange while consuming the fare of a fast food chain.

JOE: That chain's food is fit for dolts. I'll never go near it.

GAN: A four-year-old girl, perched precariously on a chair, is fiddling with a microwave oven.

JOE: This is criminally irresponsible. A deadly message for the little kids who watch. One day we'll hear of a horrible accident involving a microwave oven and an unsupervised child.

GAN: A woman is spraying the garbage can in her kitchen with some chemical supposed to annihilate its offensive odor.

JOE: Who in her right mind does that? A normal person would go instinctively for the neat and hygienic solution, taking the garbage out of the house.

GAN: A dog is shown to prefer some unappetizing-looking pellets to a slice of juicy stake.

JOE: If I had a dog like that, I would conclude that it must be mad and I'd have it put down humanely, to save canine blushes all-around.

GAN: You see a beautiful young woman using an expensive cosmetic product and saying that it will take years off you in a matter of weeks.

JOE: Why don't they show a more mature woman instead, before and after? That would really prove their point. The manufacturer's claim is baseless. I'd be wasting my money buying that product.

GAN: Two male spectators at the opera are carrying cans of beer inside their jackets.

JOE: The ad seems to say that culture is not for real men, who find opera a boring waste of time and prefer to swill beer instead.

GAN: A car salesman is pointing at various vehicles, talking excitedly about their low prices.

JOE: He should describe and praise their mechanical strength and safety features, not boast about how cheap they are. His merchandise is probably of dubious quality. Buy a car from him and regret it. You always get what you pay for.

GAN: The office supervisor in a credit card company is inflicting physical punishment on an employee when he doesn't give the customers the prescribed answer.

JOE: The company encourages bullying, persecution of the weak and vulnerable. It should be taken to court.

GAN: A family of polar bears is fraternizing with a flock of penguins, sharing soda drinks.

JOE: The unBRITEs who made the ad don't know, or ignore the fact, that polar bears live at the North Pole and penguins at the South Pole. If they brought them together deliberately, they certainly created confusion in the mind of many viewers, jumbling up what little knowledge of geography they had received in high school.

"We all need to put bread on the table," I said, "including the ad producers. Not all of them can be first class directors and scriptwriters."

"You don't have to be a genius to come up with a decent, safe, clear, to-the-point, humorous commercial. Some ads may also not tell the whole truth."

I laughed. "Ads that don't tell the whole truth! How unusual! You got any specific examples?"

"I have a JOE who tells Gan what happened when he went to see the customer relations coordinator, Coo, of a cable company about their high speed internet service."

JOE: You advertised the cost of this as $20 per month. Yet when my first bill came, it showed a total of $70. What are these extra charges? I don't understand them.

COO (looks at the paper handed to him): The first one, DUT, is the Daily Use Tax.

JOE: Why should I pay $5 for that? I don't use the connection every day.

COO: It is unwieldy to design a tax for every type of usage, so we went for a flat rate.

JOE: Why a flat rate at the highest level—daily use—and not one at the lowest level—no use at all? What's wrong with a $0 tax?

COO: We are not a charity.

JOE: Here is the next one: $10 DHT. What's that?

COO: It's the Delivery to House Tax.

JOE: You are joking, right? Where would you deliver it if not to my house? Would you take off the $10 if I asked you to move the delivery point to a tree in my backyard?

COO: I like your sense of humor.

JOE (pointing down the list): And this: $15 ERF. What is it?

COO: The Environment Responsibility Fund. We must make sure that our installations do not harm the neighborhood.

JOE: You buried your cables underground once and for all before the subdivision was built. All maintenance is done through special access points. What harm are you talking about?
COO: If a cable catches fire, we might have to dig up people's gardens to get to it and insert a replacement.
JOE: So the fund pays for the damage done to those gardens?
COO: Oh, no! The fund pays for hiring an attorney to defend our company if house owners sue us.
JOE: How about the last one—$20 WSYT?
COO (uneasy): Uhm...this is rather difficult to explain.
JOE: Let me guess: the We-Sock-it-to-You Tax?
COO (ashen-faced): I don't know what to say, sir. I need to speak to my supervisor.
JOE (in a decisive voice): Yes, you do that. And while you are at it, tell him to take your overpriced services, crook mentality, and deceitful practices and shove them all in your environmentally friendly hole in the ground.

"A JOE who pulls no punches," I said. "Although he exaggerated a little, he made his point. It's very annoying to have attractive prices dangled in front of your eyes, which become a lot less so once all the hidden extras are thrown in. Any other examples?"

"Oh, lots! There is a style of ad that puts adult words in kids' mouths. Does anybody, I wonder, find that funny? Then there are the ads about slimming diets, which exploit the desperate by making extravagant claims. And ads that sound downright moronic. Listen to the dialogue between the salesman, Sal, and a customer, Cus, in one of them.

SAL: Do you know that if you were to walk barefoot on broken glass, you'd hurt yourself very badly?
CUS (concerned): I would?
SAL: You'd get nasty cuts on your soles, which would bleed and become infected. You'd be admitted to hospital and may have both legs cut off. In extreme cases, you may be disemboweled.
CUS (alarmed): Really? So what should I do?
SAL: Wear X-brand shoes. Feet are always comfortable and safe in them.
CUS (greatly relieved): Thanks for telling me. I'll go out and buy X-brand shoes immediately, and I'll wear them all the time. I won't even take them off at night.

"Then there is the silly Cus who babbles about an item as if it were the be-all and end-all of life."

CUS (volubly): You know, for as long as I can remember, I hurt something awful every time I tried to turn my head to look between my shoulder blades, it was unbearable, I just didn't know what to do. Then my cousin Mary told me to try this product—well, I thought, what's to lose?—so I did it. You wouldn't believe how fast it worked! (Giggles meaninglessly.) The pain vanished! My head now swivels this way and that like never before, I can bite the back of my neck, such a wonderful, wonderful feeling, this product is a miracle, to be sure, and my petunias have come out great, too!

"Finally," JJ went on, "you have the guilty-by-omission ad that, by not telling the full story about what the sponsor wants, often results in wasted effort and unpleasantness. Recently a female Gan saw a TV ad urging people to apply for well-paid work with flexible hours. Curious to see what kind of work that was, she went to the given address, which turned out to be a local government office. Here is what the VOTSIT recorded of her conversation with the manageress, Man."

GAN: Can you tell me a bit more about the job?

MAN (points to a glass wall separating her office from a large hall full of women seated in front of computers): You type addresses on a standard letter, print the letters, and stuff them in envelopes.

GAN: Interesting. May I have an application form, please?

MAN (gives her the form, waits until she has finished, reads it): I'm very sorry, but we cannot employ you.

GAN: Why not?!

MAN: You are a university graduate.

GAN: Never mind that. All my life, ever since I can remember, I wanted to type letters and mail them.

MAN: This is not a job for you.

GAN: You don't think I can do it? My typing is both fast and accurate. And I could show you how to improve the efficiency of this office.

MAN (puzzled): What are you talking about?

GAN: Look, in this setup you have as many computers and printers as you have workers. I've noticed that every woman takes 9 seconds to type an address and send it to the printer, waits for 3 seconds until the letter is printed, spends another 3 seconds to fold the letter, and then another 3 seconds to put it inside an envelope, stick down the flap, and throw the closed envelope on the conveyor belt behind her. This means that she produces one ready-to-mail letter every 18 seconds. Now suppose that you change the setup and split the women into groups of 5, each group with only three computers connected to the same printer. Three women type, one woman picks up the letters as they come out of the printer and folds them up, and the last one puts the letters in envelopes. The printer queues up the letters sent by the three typists and delivers one every 3 seconds, working nonstop. Due to the continuity of the process, the new arrangement would have one ready-to-mail letter thrown on the belt every 3 seconds. As things stand at the moment, in 1 hour, or 3600 seconds, 5 typists complete $(3600 \div 18) \times 5 = 1000$ envelopes, whereas my solution would yield $3600 \div 3 = 1200$ envelopes: an improvement in productivity of 20%. To say nothing about the savings you'd make on computers and printers. So, what do you think? Will you hire me?

MAN: Definitely not.

GAN (turns on the truth-inducing function of the VOTSIT): But don't you see how useful I'd be? I'd save the government millions in expenses!

MAN: I don't care about the savings to the government. If I hired you, my boss would soon realize how much better you are at running this office and I'd be out of a job. I don't want you here, with your superior education, sharp mind, neat appearance, and professional manner. What I want is someone less intelligent than me, who I can order around; someone as inefficient and insecure as the rest of us; someone who likes spending time gossiping over coffee, making private phone calls, and playing solitaire on her screen while her work is done by an unpaid volunteer. However, since I can't write this on your application, I'll just say that you're overqualified, and nobody'll be any the wiser.

"Used as a survival weapon by unBRITE second rate plodders, the false argument of overqualification is one of the most hypocritical and discriminatory I've come across. It should've been outlawed a long time ago because all it does is slow down the progress of civilization by fostering mediocrity. Ads like this should come with a warning: 'This work environment discourages initiative and strong leadership. Outstanding candidates need not apply.'" JJ leaned back in his chair. "Now," he said, "it's time for your stories."

I finished the last of the coffee and pushed the empty cup aside. "OK, here we go. Today I've got two. They touch on mathematics and also relate indirectly to the advertising business. First one coming up."

At an army training camp, the commanding colonel is told by the higher brass that his unit has fallen badly behind in selling additional insurance to newly enlisted men, and is instructed to remedy the situation without delay. Not exactly sure how he could improve the figures, the colonel summons his noncommissioned officers and orders them to go around the mess tables every day and persuade the young men to buy life insurance policies.

A few days later the colonel checks the numbers and sees that they are pitifully low except for one particular sergeant, who has sold a very large number of policies. Intrigued, the next day he shadows the sergeant and listens to his spiel.

"Look," the sergeant says to the recruits around him, "if we go to war and you die, the army will pay your designated beneficiary 100 grand. If you buy this extra policy and you later die in combat, then the army will cough up 250 grand. Now, which soldiers do you think the army will send to the front line first: those with additional insurance, or those without?"

"That's what I call an ingenious sales pitch," JJ said appreciatively. "It shows how powerful logic can be if you have the skill to adapt it to your needs."

"And here's the second story," I offered.

A student sees an ad on TV about a new pill that, when taken, deposits in the brain all necessary knowledge in a given discipline. Since he has three tests the next day and hasn't studied enough, he decides to go to the neighborhood pharmacy to check out the claim.

"Do you have a chemistry pill?" he asks the pharmacist.

"I do." The pharmacist reaches into a little drawer and produces a tiny smooth red pill. "You swallow this and you'll know all the chemistry you need."

The student is not convinced, but pays for the pill and takes it down with a mouthful of water. Within a few seconds, his mind fills up with the atomic structure of various substances, valences, equations of chemical reactions, and so on.

"Oh, my! Amazing!" he exclaims. "Do you also have a history pill?"

"Sure do," the pharmacist says, bringing out a tiny smooth blue pill from another drawer.

The student buys this one as well and swallows it. And indeed, he suddenly knows all the names of long-dead kings, places and dates of old battles, and a lot of other such details.

"Wow!" he marvels. "This is incredible!" Emboldened, he decides to go for broke and says, "My third class test tomorrow is in mathematics. Do you have a math pill as well?"

The pharmacist scratches his head and shrugs. "I'm not sure. I might, or might not. Let me see." He rummages through all the drawers in the main room, then goes to the little store room behind him and eventually comes out with a pill in his hand. "Here, this is it. The last one."

The pill is jet-black, rough-surfaced, large as a peach stone, and covered with spikes and sharp edges all over. The student eyes it in dismay and shakes his head, "This looks awful. How am I supposed to take it down?"

"Well," the pharmacist says, "join the club: you are not the only one who finds math hard to swallow."

"Spikes or no spikes, I bet a lot of college students would give their back teeth to be able to buy such a pill." JJ put his hands on the table. "Thanks for the stories," he said. "I've really enjoyed our little chat. If all goes well, we'll talk some more next time." Then, unceremoniously, he sprang to his feet and promptly walked out of the bar.

Notes After the Meeting

Here are the answers to JJ's questions together with some brief personal comments.

Solution to Question 1 (arithmetic). The difference between the ages of the father and son remains constant at all times: $43 - 16 = 27$. First, we represent the two ages by points on the real line:

```
                Son                    Father
├────────────────┼──────────────────────┼──────────────►
0               16                     43
```

Next, we regress both ages to the time of the son's birth, when, clearly, the father was 27:

```
        Son                Father
├────────┼──────────────────┼────────────────────►
0                          27
```

Finally, we move both ages forward until the age of the son is the same as the age of the father when the son was born, that is, until the son is 27, which pushes the age of the father to 54:

```
                          Son                    Father
         |                 |                       |
         0                 27                      54
```

At this moment the father's age is twice that of his son. The event will happen when the father will be 54, in other words, after $54 - 43 = 11$ years.

Solution to Question 1 (algebraic). Let x be the number of years that need to lapse before the father is twice as old as his son. This translates into the equation
$$43 + x = 2(16 + x),$$
with the unique solution $x = 11$.

Evidently, the latter method is much simpler if you know algebra. The former is handy for someone intuitive who can operate only with numbers.

An alternative formulation of this question does not need to specify the current ages: if a father is 27 years older than his son, how old will he be/was he when his age will be/was twice that of the son? The answer is still 54.

Solution to Question 2. Going to first principles, we actually show that the given number has integral factors different from itself and 1.

This is trivial if n is even: in this case, the number $n^4 + 4^n$ is also even and, thus, not prime. So let us assume that n is odd: $n = 2k + 1$, which, for $k = 1, 2, 3, \ldots$, yields $n = 3, 5, 7, \ldots$. Using the properties of powers, we can now write the given number as

$$\begin{aligned} n^4 + 4^n &= (2k+1)^4 + 4^{2k+1} \\ &= \left((2k+1)^2\right)^2 + (2^2)^{2k+1} \\ &= \left((2k+1)^2\right)^2 + \left(2^{2k+1}\right)^2. \end{aligned}$$

This expression is of the form $a^2 + b^2$. If we complete the square on the right-hand side by adding and subtracting $2ab$—that is, $2(2k+1)^2 2^{2k+1}$—we get

$$\begin{aligned} n^4 + 4^n &= \left((2k+1)^2\right)^2 + 2(2k+1)^2 2^{2k+1} + (2^{2k+1})^2 - 2(2k+1)^2 2^{2k+1} \\ &= \left[(2k+1)^2 + 2^{2k+1}\right]^2 - 2^{2k+2}(2k+1)^2 \\ &= \left[(2k+1)^2 + 2^{2k+1}\right]^2 - \left[2^{k+1}(2k+1)\right]^2. \end{aligned}$$

Finally, using the formula $x^2 - y^2 = (x+y)(x-y)$ for factoring the difference of two squares, we obtain

$$\begin{aligned} n^4 + 4^n = &\left[(2k+1)^2 + 2^{2k+1} + 2^{k+1}(2k+1)\right] \\ &\times \left[(2k+1)^2 + 2^{2k+1} - 2^{k+1}(2k+1)\right]. \end{aligned}$$

Since, for any $k \geq 1$, each of the two factors on the right-hand side is a positive integer greater than 1, the given number is not prime.

The method of mathematical induction[4] is of little use in this case because the statement 'is not prime', while absolutely clear and rigorous, cannot be encapsulated in a crisp mathematical formula for manipulation in the inductive phase of the procedure.

A Word to the Wise

Quadratic Polynomials and Equations

Sometimes it is necessary to 'complete the square' in a quadratic polynomial. This may happen, for example, in a second semester calculus course taught in the time-honored classical fashion, where electronic computing devices are not allowed in quizzes and tests and young brains are called upon to do what they have been created for: think. This is easily done (the completion of the square, not the thinking).

 DO

Here is a simple example:
$$x^2 - 4x + 8 = (x^2 - 4x + 4) + 4 = (x-2)^2 + 4.$$

 DON'T

Completion of the square can certainly be carried out in more than one way; thus, a second possibility for the above polynomial would be

$$x^2 - 4x + 8 = \frac{1}{2}x^2 + \left(\frac{1}{2}x^2 - 4x + 8\right) = \left(\frac{1}{\sqrt{2}}x\right)^2 + \left(\frac{1}{\sqrt{2}}x - 2\sqrt{2}\right)^2.$$

But this is not helpful in calculus, where what we really aim to achieve by completing the square is to isolate the variable x inside one single term. The alternative shown here has x in both terms, which, from a calculus point of view, is undesirable.

 DO

Another handy thing to remember is the formula that yields the two solutions of the general quadratic equation. For $a \neq 0$,

$$ax^2 + bx + c = 0 \quad \Leftrightarrow \quad x_1, x_2 = \frac{-b \pm \sqrt{b^2 - 4ac}}{2a}.$$

With these solutions, we can write the quadratic polynomial decomposition

$$ax^2 + bx + c \equiv a(x - x_1)(x - x_2).$$

This is an identity, therefore, it is valid for all numbers x.

---

[4] A brief explanation of this method can be found on p. 258.

SCAM 12

Highway Driving

> *Mit der Dummheit kämpfen Götter selbst vergebens.*[1]
> Friedrich von Schiller[2]
>
> *Quem Jupiter vult perdere, dementat prius.*[3]
> Anonymous

I met JJ again at a mathematics conference sponsored by Lehigh University in Bethlehem, Pennsylvania. Just like the first time, he was sitting alone at a table in a corner of the hotel bar, gazing at me across the room with dark, inscrutable eyes.

As soon as I sat down, he pushed a piece of paper in my direction. "Here," he said laconically. "For your students."

I glanced at the paper and smiled. The text read something like this.

Question 1 (requires basic algebra). *What, if anything, is wrong with the following assertion and proof?*

Theorem. *A large part of the population is insane.*

Proof. (i) First, we show that all numbers are equal.

Let x and y be any two numbers, and let p be their arithmetic mean (average); that is,

$$p = \tfrac{1}{2}(x+y), \quad \text{or} \quad x+y = 2p.$$

This can be rewritten in the alternative forms

$$x - 2p = -y, \quad y - 2p = -x.$$

Multiplication of both sides in the first equality by x and in the second one by y yields, respectively,

$$x^2 - 2px = -yx, \quad y^2 - 2py = -xy.$$

Since the right-hand sides above are equal, it follows that so are the left-hand sides:

$$x^2 - 2px = y^2 - 2py.$$

---

[1] With stupidity the gods themselves struggle in vain. (German)
[2] German dramatist, poet, and historian, 1759–1805.
[3] As quoted by James Duport (classical English scholar, 1606–1679): whom Jupiter wishes to destroy, he first makes mad. (Latin)

We add p^2 to both sides and arrive at the equality
$$x^2 - 2px + p^2 = y^2 - 2py + p^2,$$
which we immediately recognize to be, in fact,
$$(x-p)^2 = (y-p)^2. \tag{12.1}$$
Taking the square root of these equal nonnegative numbers, we find that
$$x - p = y - p. \tag{12.2}$$
Finally, we add p to both sides to obtain $x = y$, which proves that any two numbers, therefore all numbers, are equal.

(ii) We now turn to the assertion of the theorem. Consider an arbitrary driver who might be going at, say, 105 km/h.[4] By part (i), he is also going at 300 km/h, or, indeed, at any arbitrarily large speed, which is completely insane. Since a large part of the population drives cars and other vehicles, the theorem is proved.

Question 2 (requires basic arithmetic). Estimate the thickness of a sheet of ordinary writing paper that has been folded 50 times.

"Interesting," I said. "Thanks." I folded the paper and put it in my pocket, waiting for JJ to broach, as usual, a subject of his choice. When that didn't happen immediately, I asked, just to make conversation, "Did you have a pleasant flight?"

"I didn't fly," he replied. "This time I drove. But I wish I had taken the plane."

"Why? Journey too long? Roads too busy?"

"No. Drivers too unruly."

"Oh?" I said, feigning surprise.

"What's wrong with some humans? They are perfectly decent until they get behind a steering wheel. Then, suddenly, they seem to lose all their self-control and respect for others."

"You placed bad drivers just above violent criminals on your scale, with a MICQ no higher than 4. Why such a low rating?"

"They endanger lives. It may come as a surprise to you, but JOEs don't like them at all. JOEs' views are based on simple logic:

> No good citizen ever knowingly disobeys the law.
> The highway code is part of the law.
> Therefore, a driver who does not observe the highway code is not a good citizen.

As an example, look at the findings of one of Gan's VOTSIT surveys on the subject among the population at large. He described various traffic situations to his interviewees, all involving a selfish, inconsiderate motorist, Sim, then

---

[4] Roughly, 65 mph.

used the image synthesizer to visualize their reactions. Here are some of the less extreme answers he got."

GAN: You are driving on a country road at the maximum legal speed. Sim wants to overtake you but cannot because of the traffic coming from the other direction. He blows his horn several times, trying to pressure you to move faster or get out of his way. You ignore him and concentrate on your driving. Sim guns his car and finally overtakes you on the grass verge, making obscene gestures as he passes. What do you see?

JOE: Sim's car is in a marshy ditch, a couple of miles down the road. Dazed and covered in mud, he is waiting for someone to stop and help him. I'm driving by, waving my hand at him and smiling.

GAN: You are in a long line of slow-moving cars intending to come off the highway. Sim tries to jump the queue by inserting himself in it close to the exit lane's separation point. What happens next?

JOE: A concrete barrier springs up between Sim's lane and the exit, forcing his car onto a track marked by a bright traffic sign with the inscription "London, England, via the ocean floor. Next exit and services 4,351 miles. No stopping or turning. Violators will be annihilated."

GAN: You are at a traffic light that has just changed to green. Sim is speeding along the road perpendicular to yours, disobeying his red signal. As he does so, he collides with the car in front of you, which had lawfully advanced into the intersection. How does this end?

JOE: Sim is approaching a large junction. An enormous military tank is rushing menacingly toward him from behind, showing no intention to stop. Although the light ahead is green, the junction is crossed in both transverse directions by fast-moving, bumper-to-bumper, endless convoys of huge armored trucks. A voice booms lugubriously in Sim's ear: "Payback time! Have done unto you what you do unto others!"

"Among the many responses I found two particularly interesting," JJ said. "They must've come from JOEs with good mathematical knowledge, as reflected by the retribution method they contrive. The first one shows that some people cannot cope with basic arithmetic if their life depended on it. Gan tells JOE he is driving on the highway when, without warning, Sim changes lanes, rams the back of JOE's car with great force, then accelerates and vanishes in the traffic. In JOE's immediate reaction, the steering system of Sim's car has been damaged by the impact, and he smashes straight into the massive concrete support of an overhead bridge. Fully concertinaed, the car somersaults a few times and comes to rest in the adjacent field. Sim, in searing pain and breathing a hot and fetid sulfur atmosphere, is approached by a gaunt, midnight-colored silhouette who introduces himself as the Definitive Extinguisher and Terminator, Det, and engages him in the following dialogue.

DET: This is your third strike, and you know what that means. But before taking action, I need to determine if some degree of clemency might be justified.

I'll give you a simple problem. Solve it, and you'll survive. How far did you go in school?

SIM (bewildered): Twelfth grade.

DET: Very well. Then here is an eighth grade question for you. Today, John's age exceeds the double of Mary's age by four years. Ten years ago, John was 30. How old was Mary at that time? I'll accept only the first answer, so think carefully.

SIM (after less than a second): 13!

DET: How did you come up with that number?

SIM: If we subtract 4 from 30, we find that the double of Mary's age at that time was 26, so she was 13.

DET: Really? Since John was 30 ten years ago, he is now 40. Taking away the 4-year excess leaves the double of Mary's current age as 36; therefore, Mary is 18, which means that ten years ago she was 8, not 13. You thought that the relationship between the two ages ten years ago was the same as today. You were overconfident in your mathematical ability just as you were in your driving skills. Too bad. (Prods him with a razor-sharp scythe.) C'mon, c'mon, let's go. We can't waste eternity!

"In the second of my favorite responses," JJ said, "Gan tells JOE that while he is approaching a highway exit, Sim bolts from behind, weaving dangerously from lane to lane. He doesn't signal, cuts across three lanes in one move, and gets onto the exit ramp, missing JOE's car by a whisker. In his reactive description, JOE sees the road ahead of Sim rising and curling up into a gigantic circular band of asphalt that traps Sim on its inner surface. A robed shadowy figure appears in the passenger seat and introduces himself to Sim as the Getting-even Implementation Director, Gid. Their verbal exchanges are most unusual."

GID: What kind of driving is that? Don't you have any intelligence?

SIM (indignant): I'm the holder of a bachelor of science degree!

GID: Oh, you are, are you? Very well, then. Since you've got a college education, I'll give you the opportunity to use whatever technical expertise you still possess to save your life. The radius of this road loop is 200 m. If you work out how fast you need to go to beat the gravitational force and not fall off while driving around inside it, you'll live. However, if your answer is wrong by more than 10%, you and your car are history.

SIM (petulantly): I don't remember all the formulas from Physics 101.

GID: I can help—but it's going to cost you.

SIM: How much?

GID: For every hint you solicit, you'll lose a limb.

SIM (has no choice and thinks, like all irresponsible and arrogant drivers, that he is very smart and will need at most a couple of clues to get away with his life): I accept. May I use my calculator?

GID: No. To make things fair, though, I'll stop the flow of time while you are doing your mental computations. For free.

SIM: I need to balance the gravitational force, which pulls me down, and the centrifugal force, which pushes me up, at the top of the loop. What's the expression of the former?

GID: This will cost you the left arm.

SIM (reckons he can still drive the car with the right arm alone): Okay.

GID: It is mg, where m is the mass and g the acceleration of gravity.

SIM (sees his left arm drop off): And the centrifugal force?

GID: This will cost you the right leg.

SIM (finds it hard to accept, but thinks that having to reach the gas and brake pedals with his left leg is better than being dead): Hit me.

GID: It is mv^2/r, where v is the speed and r is the radius of the loop.

SIM (sees his right leg drop off): If I equate the two opposing forces, I get

$$mg = mv^2/r;$$

if I divide through by m, I find that

$$g = v^2/r;$$

if I solve for v, I arrive at

$$v = \sqrt{gr}.$$

You said that the radius of the loop is 200; the value of g I remember from school as 2^5, or 32, so, if I replace them both in the final formula, I obtain... let's see...

$$v = \sqrt{32 \times 200} = \sqrt{6400} = 80 \text{ m/s},$$

or

$$80 \times 3600 \div 1000 = 288 \text{ km/h},$$

which is

$$288 \div 1.609 \cong 179 \text{ mph}.$$

This is impossible! My car cannot reach that kind of speed! You tricked me!

GID (unperturbed): I did not. You tricked yourself. You took g to be 32, which is its value in ft/s^2, instead of the 9.8 m/s^2 that matches the unit used for the radius. The correct result should have been

$$v = \sqrt{9.8 \times 200} = \sqrt{1960} \cong 44.3 \text{ m/s},$$

or

$$44.3 \times 3600 \div 1609 \cong 99.12 \text{ mph}.$$

Your car can do more than 100 mph. Since your error exceeds the allowed margin of 10%, you've lost the game.

"Physical units," JJ said. "A contentious subject that we should discuss in more detail sometime."

"And these are the *less* extreme answers?" I asked. "I wonder what the extreme ones look like."

"You don't want to know. As I said, ordinary people can get very angry with Sim and his kind." JJ passed a finger along the edge of the table. "Would you also be interested to hear what Sim has to say?"

"Gan VOTSITed him as well?!"

"The project wouldn't be complete without his views."

Once again, JJ's machine switched on.

GAN: Why do you always ignore the speed limit?

SIM: Driving fast is an expression of my personality. Since the Constitution guarantees me full freedom of expression, I drive fast.

GAN: Why do you never signal when changing direction?

SIM: Signaling is for little old ladies and nerds.

GAN: Why do you run red lights?

SIM: It's a challenge that makes my juices flow.

GAN: Why do you cut in front of others when there isn't enough space, forcing them to slam on the brakes or take evasive action? Don't you know that your criminal thoughtlessness can turn you into a killer?

SIM (snorts contemptuously): The roads are for the manly. The wimps should stay at home and drink tea. I'm the king of the highway. It's mine, to do on it as I please.

GAN: Your little boy rides in the front passenger seat, with the seat belt unfastened. Isn't this irresponsible, not to say illegal?

SIM: I want him to grow up fully uninhibited. Only then will he know the meaning of real freedom and become one of life's winners. The law is just a handy tool for the government to enslave spineless fuddy-duddies.

GAN: What's 5 divided by 0?

SIM: 0, of course. Why?

GAN: Just checking. As I thought, you pay no heed to the First Mathematical Commandment, so your MICQ is at most 4. You've got unBRITE written large all over you."

"Some men have dreadful road manners," I admitted.

"It's not just the men. Women do it, too. Young people, old people."

"That's human nature for you. Imperfect. You think there's a cure for this? Can we eliminate bad driving completely?"

JJ pondered. "Natural selection is helping to a certain extent: quite a few wrongdoers self-terminate. But the Darwinian[5] theory will not solve the problem by itself. The only efficient solution I can think of is to slam the guilty with what hurts them most."

"Fines?"

"Heavy fines, yes. Suspension or loss of license. Incarceration."

"For that," I said, "you need to catch them first. This means a lot more highway patrols than any state can afford to hire."

---

[5] Charles Robert Darwin: English naturalist, 1809–1882.

"Patrols are not cost effective. What's wrong with using technology? Install centrally monitored, camera-connected speed detectors at strategic points—in other words, extend the speed trap network. The states will treble their income overnight and could use the surplus for all the necessary surveillance equipment and personnel, for road maintenance, and for other worthy causes. Everybody wins, except Sim. Making stupidity pay for itself is an act of natural justice."

"It won't work," I said. "A lot of people take exception to the idea of being monitored on the road. They claim it's unconstitutional, a violation of their civil liberties. If such a measure is implemented, I can see the states using that surplus income not for road improvements, but to settle damages awarded against them in litigation."

"Why would law-abiding citizens be worried about road cameras?"

I threw a quick, oblique look over my shoulder. "The Big Brother[6] syndrome. They think it's the thin edge of the wedge. Cameras on the road today, in the office tomorrow, in the home after that. You've lived among us long enough to know better."

"Isn't this knee-jerk reaction hypocritical? Big Brother has been watching JOE for quite some time already. He's doing it through the policemen patrolling the streets, the ATM machines, the credit card payments, the details recorded on driving licenses, the surveillance cameras in shops and other public places, the gismos at the entrance and exit of toll roads which register the place and time of a car's passing, the telephone directories, the airport searches—and so on. Why don't people go to the barricades over any of these, yet they are up in arms if cameras are installed on highways? What do they think those cameras would capture apart from their number plates, and that only when they drive improperly? If they do nothing wrong, they have nothing to fear."

"Yes, but there is concern about errors and abuses."

"Abuses and malfunction happen all the time, everywhere around us," JJ said. In traffic situations, their incidence can be drastically reduced if, for example, someone fined on the basis of a camera recording is given the right of appeal in a court of law. Then the transport authorities would have to produce the evidence to prove their case."

"That would lead to unmanageable congestion in courts."

"Not if the law imposes a doubling of the fine when the plaintiff loses. Drivers who know they were in the wrong would soon learn to pay the fine instead of appealing."

"More controversy," I noted.

JJ shook his head slowly. "Perhaps you should start with simpler measures. For example, don't allow car manufacturers to make vehicles capable of speeds above the legal limit. Or raise the driving age to 20."

"Are you serious? That wouldn't fly for a second! The great majority of

---

[6] The nickname given by George Orwell (see footnote 2 on p. 101) in his book *Nineteen Eighty-Four* to the repressive leader (dictator) of the fictitious totalitarian state of Oceania.

parents can hardly wait for their kids to turn 16 so they don't have to ferry them around any more."

"I find this highly illogical: you don't trust teenagers to drink beer, yet you allow them to drive a car on public roads?"

"Youngsters feel indestructible and are tempted to do things to excess," I said.

"I agree. But drinking too much alcohol harms only their own body. Driving too fast imperils others. You must also be more severe with the motorists responsible for fatal accidents. The courts ought to give them a real wallop, not just a slap on the wrist. Recklessness leading to loss of life or limb should be regarded as an aggravating circumstance."

"That happens sometimes."

"Yes, once in a blue moon." JJ paused briefly. "And what about your driver's permit test?" he continued. "When I took it, the theoretical part was a joke. I was shown to a computer, told to press a key, read the text on the screen, and choose one of four given answers. A typical question sounded something like this."

> You are approaching an intersection. Which of the following should you do?
> A. Close your eyes, accelerate like mad and drive through.
> B. Get out of the car, walk into the junction, signal with your hands to everyone else to stop, then go back to the car and drive across.
> C. Drive onto the roof of the car in front of you, wait until it has cleared the intersection, then dismount and continue on your way.
> D. Slow down, look both right and left to make sure that the road you are about to cross is clear, and proceed with caution.

"I answered all 20 questions in about 30 seconds. In front of me, at another computer, was someone who did not manage one single question in that time. I saw him later, after the road test: he, too, got his license!"

"This type of testing was replaced a while ago by a proper written exam."

"As it should've been. Anyway, it's clear that Earth's MICQ will not rise above 50 in a hurry. There'll be bad drivers on your planet for a lot longer than I thought." JJ leaned back in his chair. "Now," he said, "it's time for your stories."

I finished the last of the coffee and pushed the empty cup aside. "OK, here we go. Today I've got three, all involving drivers and some interesting applications of mathematicians' main reasoning tool: logic. First story coming up."

One cold winter morning a foreign businessman arrives in Moscow and hails a taxi to go to the hotel. The driver, who has only a smattering of English, looks at the address written down in Russian on the visitor's card, nods knowingly, and takes off. The traffic is very heavy and the road surface treacherous, to say the least.

As the car approaches the first main intersection, the lights are turning yellow. The driver floors the gas pedal and flies through the junction. Taken aback by such imprudence, the visitor tries to remonstrate with him. "Didn't you see?" he asks. "The light was turning yellow! Why didn't you stop?"

The driver looks at the passenger in the rear view mirror and, pointing an emphatic finger at himself, says, *"Ya—dzhigit!"*[7]

The visitor shakes his head and, as the car swerves this way and that, curses his bad luck. When they approach the second major intersection, the lights are turning red. The driver again pushes the gas pedal all the way, sending the taxi like a bullet across the junction.

"Hey!!" the passenger yells. "Are you crazy?? Why did you run the light?"

Once more, the driver smiles and points to himself with pride: *"Ya—dzhigit!"*

The visitor, lost for words, swallows hard, grabs hold of the edges of his seat and holds on for dear life. Eventually, the taxi comes to a third intersection, and this time the lights are turning green. Clenching his jaws, the driver puts all his weight on the brake pedal and brings the car to a screeching halt.

Completely bewildered, the passenger asks, "Why did you stop now? The light just went green!"

Calmly, the driver turns his head and sweeps one arm from right to left and back, indicating the street perpendicular to their direction. "Sure," he says, "light go green. But what if another *dzhigit* is coming the other way?"

"What is the origin of the word *dzhigit?*" JJ asked.

"A friend tells me that it's Georgian and means a good horseman—a kind of cowboy."

JJ shrugged. "He may be right. Myself, however, I'm inclined to believe that the word actually originated in this country."

"The second story is also about a traffic incident," I said.

A little old lady is stopped by a policeman on the highway at the edge of a town. As the officer approaches from behind she lowers the window and asks, "Anything the matter, sir?"

"Are you aware that you were speeding, ma'am?"

"I don't think I was."

"Yes, you were. You were doing 85 in a 55 zone. May I see your driver license, please?"

"I'm sorry, officer, but I couldn't possibly show it to you."

"Why not?"

"It was taken away from me two years ago, for driving under the influence."

The patrolman frowns. "May I see your car documents, then?"

"I must apologize again, officer, but this is not my car. In fact, I killed the owner of the car at a gas station about 50 miles back, hacked his body into pieces, and stuffed them in the trunk."

At this point the patrolman says sternly, "Please remain in your car, ma'am, and don't do anything. Just wait there." His eyes on the little old lady, he walks backwards to his car and radios for help.

In a couple of minutes, four police cars arrive and the police chief himself gets out of one and cautiously approaches the woman.

---

[7] I am a macho man. (Russian)

"Ma'am," he says, pointing to the officer who had pulled her over, "my man over there tells me that you lost your license for driving under the influence. Is this true?"

"Absolutely not, sir," the lady answers calmly, rummaging through her purse. "Here it is."

The chief takes the license, examines it, finds nothing wrong, and gives it back. "That's very odd," he says. "My man over there also tells me that this car is not yours, that you butchered its owner and put his body in the trunk."

The old woman smiles and hands over a piece of paper. "See for yourself, chief: the car belongs to me. A body in the trunk? Well, have a look." She touches a lever and the trunk opens for inspection: clean and empty.

The chief, extremely embarrassed, says, "I don't understand, ma'am.... My man over there—"

"Sir," the woman interrupts, "your man over there is obviously a liar. For all I know, he may also have told you that I was speeding."

"Funny, but I hope that in real life the little old lady would get a hefty fine," JJ commented.

"The third story tries to make a different kind of point."

While a woman is frying eggs for breakfast, her husband rushes into the kitchen and shouts, "Careful! You need more oil! Don't you see?? They are going to stick! More oil! You put too many of eggs in the pan! Oh, dear God!! You never listen to me while you're cooking! Careful, I said! Turn the eggs over! Now!! Come on!!! What are you doing?? Are you out of your mind? Sprinkle some salt on them! Don't forget the salt! You never use salt! Get the salt! Careful!!"

The woman looks at her husband. "What's come over you? You think I can't fry a few eggs?"

"I know you can, dear," the man replies in a normal voice, "but I wanted to show you how it feels when I'm driving with you in the car."

"Now, now; this is bound to ruffle a few feathers." JJ put his hands on the table. "Thanks for the stories," he said. "I've really enjoyed our little chat. If all goes well, we'll talk some more next time." Then, unceremoniously, he sprang to his feet and promptly walked out of the bar.

Notes After the Meeting

Here are the answers to JJ's questions together with some brief personal comments.

Solution to Question 1. Equality (12.2) is incorrect. Recalling that

$$\sqrt{a^2} = |a|,$$

or the absolute value of a, defined to be

$$|a| = \begin{cases} a & \text{if } a \geq 0, \\ -a & \text{if } a < 0, \end{cases}$$

we see that (12.1) yields
$$|x - p| = |y - p|,$$
from which
$$x - p = y - p \quad \text{or} \quad x - p = -(y - p).$$
The former leads to the absurdity $x = y$. The legitimate choice is the latter, because it is equivalent to our original assumption:
$$x - p = -y + p \quad \Leftrightarrow \quad x + y = 2p.$$

In light of JJ's comments, the assertion of the theorem would still hold if the quantification 'A large part' were replaced by the more moderate 'Some'.

The *léger de main*[8] in this 'proof' is building up the squares of $x - p$ and $y - p$ on the sly and then conveniently choosing the wrong sign when taking the square root.

Sometimes it is necessary to square both sides of a given equation to facilitate its solution, but all the values of the unknown obtained in this way must then be tried on the original equation to eliminate the spurious ones, introduced by the squaring process. For example, the equation
$$\sqrt{x} = 2x - 1$$
is easily solved by squaring both sides; this yields
$$x = (2x - 1)^2$$
$$\Leftrightarrow \quad x = 4x^2 - 4x + 1$$
$$\Leftrightarrow \quad 4x^2 - 5x + 1 = 0$$
$$\Leftrightarrow \quad x = 1 \quad \text{or} \quad x = \tfrac{1}{4}.$$

Of these two values, only $x = 1$ satisfies the original equation; the second one leads to the impossibility $\tfrac{1}{2} = -\tfrac{1}{2}$ and must be discarded.

Solution to Question 2. A sheet of ordinary writing paper is roughly 0.1 mm, or 0.0001 m, thick. If we work in ideal conditions—that is, the fold is always perfect and does not generate additional 'bulging'—each folding operation doubles the preceding thickness, so, after 50 steps, the final thickness is
$$h = 0.0001 \times 2^{50} = 0.0001 \times \underbrace{(2 \times 2 \times \cdots \times 2)}_{50 \text{ times}}$$
$$= 112,589,990,684.2624 \text{ m} \cong 112,590,000 \text{ km}.$$

This is approximately three-quarters of the distance from Earth to the Sun, or 295.5 times the distance between Earth and the Moon.

---

[8] Sleight of hand. (French)

The operative word here is 'doubles', which places the problem firmly in a geometric progression setting. Surprisingly few people notice this, hence the large number of estimates that are widely off the mark. Even fewer realize the size of paper necessary to conduct the experiment physically. Assuming, for simplicity, that
   (i) the original sheet is square-shaped;
  (ii) the folding is always done along the line of symmetry parallel to a pair of opposite sides so that the shape alternates from square to rectangle to square and so on;
 (iii) the final shape is a square of side 1 cm (10 mm),
we work backwards to find that the original sheet has to be a square of side

$$10 \text{ mm} \times 2^{25} = 335{,}544{,}320 \text{ mm} \cong 335.5 \text{ km};$$

therefore, of approximate area

$$335.5 \times 335.5 \cong 112{,}560 \text{ km}^2,$$

which is slightly larger than the area of the state of Virginia.

An exponential function with base greater than 1 grows awfully fast.

A Word to the Wise

Square Roots

☺ DO

(i) In calculus, $\sqrt{x}$ denotes the positive square root of x; the negative square root is denoted by $-\sqrt{x}$. Consequently,

$$\sqrt{a^2} = |a|,$$

which is correct, according to the above convention, for any real number a.

(ii) The square root function does not behave linearly. Applied to a product, it is equal to the product of the square roots computed for each factor (if each of the factors has a square root); it has no such effect on a sum of terms:

$$\sqrt{a^2(x^2 + y^2)} = \sqrt{a^2}\sqrt{x^2 + y^2} = |a|\sqrt{x^2 + y^2}.$$

☠ DON'T

Woefully incorrect handling like this is seen from time to time:

☠ $\sqrt{a^2(x^2+y^2)} = \sqrt{a^2}\sqrt{x^2+y^2} = a\sqrt{x^2+y^2}$
☠ $\phantom{\sqrt{a^2(x^2+y^2)}} = a(\sqrt{x^2} + \sqrt{y^2}) = a(x+y).$

The modulus bars are missing and statement (ii) above has been violated.

SCAM 13

Units of Measurement

> *If God had intended us to use the metric system, he would have given us 10 fingers and 10 toes.*
>
> Anonymous

> *Thus the metric system did not really catch on in the States, unless you count the increasing popularity of the nine-millimeter bullet.*
>
> David Barry, Jr.[1]

I met JJ again at a mathematics conference sponsored by the Massachusetts Institute of Technology in Cambridge, Massachusetts. Just like the first time, he was sitting alone at a table in a corner of the hotel bar, gazing at me across the room with dark, inscrutable eyes.

As soon as I sat down, he pushed a piece of paper in my direction. "Here," he said laconically. "For your students."

I glanced at the paper and smiled. The text read something like this.

Question 1 (requires basic arithmetic). *What, if anything, is wrong with the following assertion and proof?*

Theorem. *Money has no value.*

Proof. *Consider the simple chain of equalities*

$$\$1 = 100¢ = 10¢ \times 10¢ = \$0.1 \times \$0.1 = \$0.01 = 1¢.$$

Based on this, we can now write consecutively

$$\$100 = 100 \times \$1 = 100 \times 1¢ = 100¢ = \$1 = 1¢,$$
$$\$10,000 = 100 \times \$100 = 100 \times 1¢ = \$1 = 1¢,$$
$$\vdots$$

from which we conclude that any sum of money, however large, is worth only 1 cent and thus, practically, has no value.

Question 2 (requires functional analysis). *What, if anything, is wrong with the following assertion and proof?*

Theorem. *The universe does not exist.*

Proof. *We will show this in four steps.*

---

[1] American humorist, b1947.

(i) First, we claim that $\sqrt{2}$ is a rational number. Consider a square with sides of unit length and construct the sequence $S_1, S_2, S_3, \ldots$ of polygonal lines from A to B consisting of the progressively thinner segments shown in Fig. 13.1.

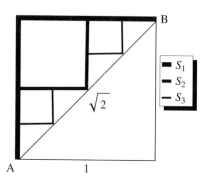

Fig. 13.1.

It is obvious from the construction that, for every $n = 1, 2, \ldots$, the total length $l(S_n)$ of each polygonal line S_n is 2. It is also obvious that, as n increases indefinitely, the polygonal line S_n gets arbitrarily close to the diagonal $S = AB$ of the square. This limiting process is written symbolically in the form

$$\lim_{n \to \infty} S_n = S. \tag{13.1}$$

The length $l(S)$ of the diagonal S is computed by Pythagoras's theorem:

$$l^2(S) = 1^2 + 1^2 = 2,$$

so

$$l(S) = \sqrt{2}.$$

Since, by (13.1), the limiting line for the sequence S_n is S, we deduce that the limiting value of $l(S_n)$ is $l(S)$:

$$\lim_{n \to \infty} l(S_n) = l(\lim_{n \to \infty} S_n) = l(S) = \sqrt{2}. \tag{13.2}$$

On the other hand, as the lines S_n approach the diagonal S, their lengths $l(S_n)$, which are all equal to 2, have the obvious limiting value

$$\lim_{n \to \infty} l(S_n) = \lim_{n \to \infty} 2 = 2. \tag{13.3}$$

Equating the values of $\lim_{n \to \infty} l(S_n)$ computed in (13.2) and (13.3), we find that

$$\sqrt{2} = 2; \tag{13.4}$$

that is, $\sqrt{2}$ is a rational number (in fact, a positive integer).

(ii) Next, we claim that $1 = 0$. *We write (13.4) as*

$$\sqrt{2} = 2 = \left(\sqrt{2}\right)^2$$

and divide both ends through $\sqrt{2}$ *to obtain* $1 = \sqrt{2}$. *Since, by (13.4), we also have* $\sqrt{2} = 2$, *it follows that* $2 = 1$. *Subtracting 1 from both sides, we arrive at*

$$1 = 0. \tag{13.5}$$

(iii) We now claim that zero is the only number. Indeed, let x *be any number. Multiplying both sides of equality (13.5) by* x, *we conclude that* $x = 0$; *in other words, any number coincides with* 0, *which is, therefore, the only number.*

(iv) Finally, we prove the assertion of the theorem. Consider an arbitrarily large, but finite, part of the universe, and let D *and* N *be its diameter and the total number of atoms of matter in it. By (iii),* $D = N = 0$, *so the universe reduces to a single massless point, therefore, to all intents and purposes, it does not exist.*

"Interesting," I said. "Thanks." I folded the paper and put it in my pocket, waiting for JJ to broach, as usual, a subject of his choice. He needed no special invitation.

"It's illogical," he began abruptly, "to see the most powerful country in the world persisting in the use of feet, pounds, pints, and the like when measuring physical quantities. These units go back to the Paleolithic days, when humans needed to relate size to parts of their bodies for comparison. Why are you so opposed to moving over to the decimal system?"

"I don't see a problem with the old units, and I don't see the advantage of the decimal ones."

"No? Think about it: why is a foot the length it is? A foot's length differs from person to person. Whose foot was used as the standard? Clearly, by the size of it, it was a man's. Yet the feminist lobby is keeping unusually quiet on this blatant piece of male chauvinism. The meter, on the other hand, is defined in the SI[2] in terms of a universal physical constant: the speed of light. Very objective, very scientific, and gender-unbiased."

"I don't think it matters so much," I said. "After all, at the end of the day they are all numbers. Do you believe that some numbers should have special privileges?"

"Not when they are just that: abstract numbers. But it's a different story when they represent measurements. Look at this conversion table for some of the units used in America." As he said that, he pushed a second piece of paper across the table. It listed the old tiresome conversion factors known practically nowhere else in the civilized world.

---

[2] *Système International.* (French)

12 inches	=	1 foot
3 feet	=	1 yard
220 yards	=	1 furlong
8 furlongs	=	1 mile
4840 square yards	=	1 acre
640 acres	=	1 square mile
16 fluid ounces	=	1 pint
8 pints	=	1 gallon
437.5 grains	=	1 ounce
16 ounces	=	1 pound

"Now," JJ said, flipping the paper over, "look at the conversion table for the units in the metric system."

10 millimeters	=	1 centimeter
100 centimeters	=	1 meter
1000 meters	=	1 kilometer
100 square meters	=	1 are
100 ares	=	1 hectare
10 milliliters	=	1 centiliter
100 centiliters	=	1 liter
1000 grams	=	1 kilogram
1000 kilograms	=	1 tonne

"Mess versus neatness," JJ said. "Bizarre versus logical. Sure, 3, 8, 12, 16, 220, 437.5, 640, and 4840 are perfectly acceptable numbers, but they don't have the ease of manipulation afforded by 10, 100, and 1000. The superior practicality of the metric system has even deeper implications. What's your smallest unit of length? The inch, right? Then, to go smaller, you use fractions of it: halves, quarters, eights, and so on. This forces people to do sums like $\frac{3}{4} + \frac{7}{16}$ in inches instead of $19 + 11$ in millimeters. No prizes given for guessing which one is easier for an elementary school student, or even an educated adult. Fractions will never be as popular as natural numbers. To recall a remark I made during one of our earlier meetings, it's small wonder that in this country kids insist on going to college whether they have the necessary ability for academic work or not, instead of wanting to earn good money by becoming, say, plumbers, carpenters, or electricians. It may also explain in part why some of your plumbers, carpenters, and electricians are guilty of shoddy workmanship."

JJ drew in a deep breath. "Another important aspect," he resumed, "is the advantages that the adoption of the metric system would bring to your economy. You wouldn't have to design two technological processes, one to produce for the internal market and the other for export. The costs would be much lower if a single set of specifications were needed in both. Not only would you export more, but, for exactly the same reason, your imports would be cheaper. People in countries outside the U.S. learn English to improve their chances when they face you in the global competition game. You should do the same by learning their technological alphabet—the metric system."

"They tried to introduce it in our schools some years back. The project was a big flop."

"They didn't try hard enough, or they used the wrong method. They should have looked at the British and learned something from their experiences. When the old pound–shilling–penny currency system was changed there to the new, decimal, pound–pence system, the change was sudden: overnight, the old currency simply disappeared, so the populace was compelled to get used to the new one because this new one was the *only* one available. On the other hand, when they wanted to change from degrees Fahrenheit—another uninspired choice of units in this country, too—to degrees Celsius, they gave the weather forecast on television with the Celsius temperature printed on the map but with the Fahrenheit one also mentioned verbally by the presenter. Of course, the viewers paid no attention to the centigrades and listened instead to the numbers they were used to. Even today there are people in Britain who talk about degrees Fahrenheit, but nobody mentions shillings any more."

"I heard that there was a lot of resistance in the U.K. when the authorities started the process of conversion to metric. People were complaining, for example, that food recipes written in grams instead of ounces would dramatically change the taste of the dishes."

"Given the reputation of British cuisine, any change brought about by the conversion could only be for the better," JJ said. "The lesson is that if you want to move from one system of units to another, you must do it suddenly and completely, making it impossible for ordinary folk to use the old tool. Accepting both systems side by side may have dire consequences. Remember the failed Mars Polar Lander? The contractor had employed the archaic units of measurement for navigation; the Jet Propulsion Laboratory engineers thought they were metric. Oh, dear me."

"In spite of the cumbersome nature of our unit system," I countered, "we did manage to put men on the Moon."

"You'd have probably done it sooner and less expensively if you had used the metric system. You really need to get rid of those silly numerical dinosaurs, which are nothing but a pathetic vestige of the old colonial times."

I didn't want to lose the argument, but, short on inspiration, I came up with something that sounded very much like clutching at straws. "Many of our units are not the same as those in the British imperial system. The British pint, for example, has 20 fluid ounces; ours has only 16. Except in our British-style pubs where they serve British and Irish beers."

"The customer is cheated in those pubs," JJ said.

"What do you mean, cheated? They use the same pint glasses as in the U.K."

"I've been to some of them, and they don't. Their glasses are a fraud: they hold exactly 20 fl oz if you fill them *to the brim*. When beer is poured into the glass, the froth takes up a volume of about 2 fl oz at the top, which means that the customer gets short-changed. By contrast, British pint glasses are larger and have a mark to indicate the 20 fl oz level. The froth is always above that mark, so no cheating there. Chalk up another one for the Brits. If only they learned how to chill beer properly."

"Well," I said, "at least, unlike them, we drive on the correct side of the road."

"Wrong again. Historically, driving started everywhere on the left, because it was more convenient and safer. The great majority of the population is right-handed, and a right-handed person finds it easier to mount and dismount a horse from its left side, and from the curb, not from the middle of the road. Thus, the near curb side had to be on the horse's left. This was even more imperative when the person was a man carrying a sword (which was worn on the left if the man was right-handed). During the 1789 French Revolution, driving on the left was regarded as an aristocratic habit, so everybody moved to the right. This was made a rule in Paris five years later. Then came Napoleon, who enforced the new rule in all the lands he conquered. They did not include the U.K. and some other parts of Europe, which continued to drive on the left. For pragmatic reasons, one by one, those other parts began changing to the right. The insular U.K. did not. Nor did U.K.'s former colonies. In the U.S. the change happened gradually after independence, as a symbol of breaking with the past. Unfortunately, the change did not include the old units of measurement, which is a pity."

"Perhaps..." I said, hoping to take the discussion in a different direction. But JJ was not done yet.

"Do you know that American engineers continue to refer to these units as *English*? It's so ironic! And their mischief-making doesn't stop there. Some of the things they do with units are downright sinister. Petroleum engineers, for example, define a certain unit for the specific gravity of hydrocarbons by means of a formula that is dimensionally inconsistent, just because they want the result to be expressed as a positive integer between 10 and 60. Another anomaly concerns their viscosity units, which have no formula for conversion to metric. Their conversion to real physical units can be done only by means of specially drawn charts. Also, engineers are guilty of using empirical relationships that mix English with metric units. I could go on and on."

"Will we ever adopt metrication, I wonder?"

"Yes, you will. A first laudable step has already been taken by the BRITEs in the medical services. When I visited a human colleague in hospital last month, I noticed that their technicians took the patients' vitals in meters, kilos, and degrees Celsius. The dam of illogical resistance to metrication has been breached, and it's just a matter of time before the power of simple and convenient arithmetic comes flooding through and wipes off the inertia of anachronistic tradition. The truth of the Ninth Mathematical Commandment will prevail, pushing a lot of humans upwards from the 40–44 MICQ band." JJ leaned back in his chair. "Now," he said, "it's time for your stories."

I finished the last of the coffee and pushed the empty cup aside. "OK, here we go. Although measurement is not an easy topic, I think I remember a couple that relate to it. First one coming up."

Two engineers are trying to measure the height of a flagpole—unsuccessfully: they have only a measuring tape, which refuses to remain vertical along the pole, bending and falling to the ground every time they try.

A mathematician passes by and asks what the problem is. "That's easy," he nods when the engineers tell him. He takes the pole from its socket in the ground, lays it down, and measures it without difficulty.

After he's gone, one of the engineers says contemptuously to the other, "So typical of a mathematician! We need to find the height, and he gives us the length instead."

"Hey!" JJ exclaimed. "An engineer would say that you got this one backwards!"

"He probably would, but he'd be wrong on two counts. First off, mathematicians never bother to know how tall a flagpole is. And, second, if they did, they would measure the length of its shadow on the ground and the angle of the sun in the sky and then use simple trigonometry to find the answer."

"What if it's a cloudy day?"

I gave JJ a stern look. "Why are you batting for the engineers? Do you want the second story or not?"

"Yes, please," he answered evenly.

"Fine. But we'll have to go to Europe for it."

A French driver, caught by the police radar driving at 250 km/h on a road with a maximum allowed speed of 70, is pleading his case before the judge.

"Your honor, I don't deny that I saw the traffic sign showing the number 70 within a red circle, but the sign did not specify any speed units. As you no doubt know, the law of July 4, 1837, proclaims the metric system to be the official unit system in France, while Ordinance 65–501 of May 3, 1961, modified in accordance with European directives, defines the SI units as the only legal units in our country. Well, the SI units for length and time are, respectively, the meter and the second. Consequently, speed is legally measured in meters per second. I cannot imagine that the ministry of justice would not uphold the laws of the Republic. So the 70 I saw must mean 70 m/s, which is the same as 252 km/h. The policeman who stopped me claims that I was driving at 250 km/h, which I fully accept. But this speed is 2 km/h below the allowed legal limit. Accordingly, I respectfully request that my driver license be reinstated and the fine imposed on me canceled."

"An ingenious defense," JJ said. "However, the courts are a law unto themselves, and anyone who pleads like this would not have a chance. Which is just as well: you don't want to see motorists driving at over 155 mph on roads with a maximum speed of 40 mph."[3]

---

[3] Approximate values for those given above in km/h.

"Today I have a bonus for you: a third story," I said. "Not related to measurement, but to students. Here it is."

GAWD is curious about how students prepare for their finals. Since he is rather busy, he summons the Gatekeeper and says unto him, "Choose a typical university on Earth, observe what the student population is doing in the run up to exams, then come back and report to me."

The Gatekeeper nods and goes away. The next day he returns. "GAWD," he says, "I've seen a sight of wonder. One week before the finals, the medical students are poring over their books, learning by heart the names of various parts of the human body, drawing sketches, mixing up chemical substances, and so on; the law students do not even sleep, spending all their time memorizing cases and articles of law and mumbling to each other in a strange jargon; by contrast, the science and engineering students do nothing but party all day and night, having a good time."

"I don't understand," GAWD says. "Why should they behave so differently? After all, I created the lot of them in my own image. Hmm ... go back and observe some more, and tell me tomorrow if there is any change."

The Gatekeeper dutifully does GAWD's bidding every day, and every day his report is the same. Until, that is, the morning before the exams, when he returns and says, "GAWD, a substantial change has occurred. While the medical and law students are working their tails off, as purposeful as ever, the science and engineering students, to the last man and woman, are down on their knees, praying hard."

"Ah!" GAWD smiles with deep satisfaction. "These are our kind of people! They are the ones we will pass tomorrow!"

But the Gatekeeper looks uncertain. "It won't be easy, GAWD. I think B.L.Z. Bub got there ahead of us. He calls himself 'the professor' and claims to be you."

"One that the students undoubtedly like to hear. In fact, some might even believe it." JJ put his hands on the table. "Thanks for the stories," he said. "I've really enjoyed our little chat. If all goes well, we'll talk some more next time." Then, unceremoniously, he sprang to his feet and promptly walked out of the bar.

Notes After the Meeting

Here are the answers to JJ's questions together with some brief personal comments.

Solution to Question 1. It is clear that the units are used incorrectly. The proper chain of equalities is

$$\$1 = 100\cent = 10 \times 10\cent = 10 \times \$0.1 = \$1,$$

which is obvious.

Although this is a rather transparent 'trick', not everybody seems to see the glaring error on which it is based.

Solution to Question 2. The crux of the matter here is how we measure the distance between various mathematical objects. Even defining the distance between two points in a plane is not always evident. Suppose that these points represent houses on a city map. You may define the distance between them to be the length of the straight line segment joining the points. But this would be useless if you were a cab driver. In that case, the distance would be better defined as the length of the shortest street route from one point to the other. For more abstract objects—for example, 'smooth' curves—the concept of distance is replaced by a generalized version known as metric, which is one of the topics thoroughly investigated in functional analysis.

Warning: a basic understanding of what it means for something called a functional to be continuous on something called a normed space would be very helpful to the reader interested in following this discussion; persons of a nervous disposition should skip the rest of the solution.

There are two essential issues with the argument in the first part of the 'proof', specifically, with (13.1) and (13.2).

(a) What does
$$\lim_{n\to\infty} S_n = S$$
really mean?

(b) Are the equalities
$$\lim_{n\to\infty} l(S_n) = l(\lim_{n\to\infty} S_n) = l(S) = \sqrt{2}$$
correct? If not, why not?

We examine these issues in more detail.

(a) The way it is written, (13.1) is an imprecise statement. Here S_n and S are geometric objects, which does not sit well with the limiting process involved. To make things rigorous, we rotate Fig. 13.1 until the diagonal AB lies on the x-axis, and think of $S_1, S_2, S_3, \ldots,$ and S as the graphs of continuous functions defined on the closed interval $\left[0, \sqrt{2}\right]$ (see Fig. 13.2).

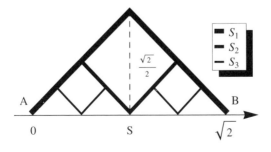

Fig. 13.2.

We write the equations of these functions, for simplicity, in the form

$$y = S_1(x), \quad y = S_2(x), \quad y = S_3(x), \quad \ldots, \quad y = S(x) \equiv 0,$$

and, acting on intuition, define the 'distance' $\|S_n\|_0$ between the lines S_n and S to be the longest vertical distance of any point on S_n from the x-axis. As is easily seen,

$$\|S_n\|_0 = \max_{x \in [0, \sqrt{2}]} |S_n(x)| = \frac{\sqrt{2}}{2^n}. \tag{13.6}$$

The number $\|S_n\|_0$ is called the norm of S_n. Since the right-hand side in (13.6) approaches 0 as n becomes arbitrarily large, we have

$$\lim_{n \to \infty} \|S_n\|_0 = 0. \tag{13.7}$$

This is the common sense interpretation of (13.1).

(b) If we accept the validity of (13.2), then, as seen in (i) and (ii) (and other 'results' that we did not pursue), all mathematical hell breaks loose. Hence, (13.2) has to be false. But what exactly is wrong with it?

The first equality in (13.2) assumes that the functional l which yields the length of the arc of a 'smooth' curve of equation $y = y(x)$ between its points corresponding to $x = 0$ and $x = \sqrt{2}$, is continuous. This functional is defined by

$$l(y) = \int_0^{\sqrt{2}} \sqrt{1 + [y'(x)]^2} \, dx$$

and can be applied to $y = S_n(x)$ because $S_n(x)$ is continuous and piecewise continuously differentiable on $[0, \sqrt{2}]$, with

$$|S_n'(x)| = 1 \tag{13.8}$$

at all x where the derivative $S_n'(x)$ is defined. Since l implicates the derivative of the function whereas $\|\cdot\|_0$ does not, it is fairly straightforward to verify that l is not continuous in the norm $\|\cdot\|_0$. Hence, if (13.1) is interpreted in the sense of (13.7), then the first equality in (13.2) is invalid. This explains the anomalous conclusion (i) and, implicitly, the ridiculous assertion of the theorem.

If we change the definition of the 'distance' between S_n and the x-axis to

$$\|S_n\|_1 = \max_{x \in [0, \sqrt{2}]} |S_n(x)| + \max_{x \in [0, \sqrt{2}]} |S_n'(x)|$$

(where $|S_n'(x)|$ has been extended by continuity to be equal to 1 also at the finitely many points where $S_n'(x)$ does not exit), we can show that the functional

l is continuous with respect to the new norm $\|\cdot\|_1$, so the first equality in (13.2) holds. But now the second one doesn't, since (13.1), interpreted with $\|\cdot\|_1$, fails: by (13.6) and (13.8),

$$\|S_n\|_1 = \frac{\sqrt{2}}{2^n} + 1,$$

therefore,

$$\lim_{n\to\infty} \|S_n\|_0 = 1 \neq 0.$$

This question shows that some simple-looking mathematical chicaneries may require quite intricate and sophisticated explanations.

A Word to the Wise

Factoring and Roots of Equations

For many purposes, including simplification of fractions and solution of algebraic equations, factoring complicated expressions is very useful.

☺ DO

Here is an elementary example of fraction simplification. We assume that the denominator on the left-hand side is nonzero.

$$\frac{a^2 + 3ab + 2b^2}{a^2 + ab - 3a - 3b} = \frac{(a^2 + ab) + (2ab + 2b^2)}{(a^2 + ab) - (3a + 3b)}$$
$$= \frac{a(a + b) + 2b(a + b)}{a(a + b) - 3(a + b)}$$
$$= \frac{(a + b)(a + 2b)}{(a + b)(a - 3)}$$
$$= \frac{a + 2b}{a - 3}.$$

 DON'T

Someone with little knowledge of algebra might write

$$\frac{a^2 + 3ab + 2b^2}{a^2 + ab - 3a - 3b} = \frac{3ab + 2b^2}{ab - 3a - 3b} \quad (a^2 \text{ was 'canceled'})$$
$$= \frac{3 + 2b^2}{-3a - 3b} \quad (ab \text{ was 'canceled'})$$
$$= \frac{2b^2}{-a - 3b} \quad (3 \text{ was 'canceled'})$$
$$= \frac{2b}{-a - 3}. \quad (b \text{ was 'canceled'})$$

☺ DO

Factoring is all-important when solving equations. First, all terms should be shifted to the left-hand side, then that side should be fully factored and each factor should be set equal to zero to harvest all possible roots. This is illustrated below by means of a simple algebraic equation:

$$x^2 - 3x = -2$$
$$\Leftrightarrow x^2 - 3x + 2 = 0$$
$$\Leftrightarrow (x-1)(x-2) = 0$$
$$\Leftrightarrow x - 1 = 0 \quad \text{or} \quad x - 2 = 0$$
$$\Leftrightarrow x = 1 \quad \text{or} \quad x = 2.$$

☠ DON'T

The following kind of 'manipulation' has no justification whatsoever:

$$x^2 - 3x = -2$$
☠ $\Rightarrow x(x-3) = -2$
☠ $\Rightarrow x = -2 \quad \text{or} \quad x - 3 = -2$
☠ $\Rightarrow x = -2 \quad \text{or} \quad x = 1.$

Only when a product of factors is equal to zero can one apply the factor-by-factor argument.

SCAM 14

Percentages and Living on Debt

> *Forgetfulness: a gift of God bestowed upon debtors in compensation for their destitution of conscience.*
>
> Ambrose Bierce[1]
>
> *I can get no remedy against this consumption of the purse; borrowing only lingers and lingers it out, but the disease is incurable.*
>
> William Shakespeare[2]

I met JJ again at a mathematics conference sponsored by Northwestern University in Evanston, Illinois. Just like the first time, he was sitting alone at a table in a corner of the hotel bar, gazing at me across the room with dark, inscrutable eyes.

As soon as I sat down, he pushed a piece of paper in my direction. "Here," he said laconically. "For your students."

I glanced at the paper and smiled. The text read something like this.

Question 1 (requires basic algebra). *What, if anything, is wrong with the following assertion and proof?*

Theorem. *Earth is flat.*

Proof. (i) First, we demonstrate that $1 = 2$. We start with the obvious equality

$$a^2 - a^2 = a^2 - a^2,$$

where a is any nonzero number. We will deal with the two sides differently: on the left we factor out a, whereas on the right we use the formula for factoring the difference of two squares, namely,

$$x^2 - y^2 = (x+y)(x-y);$$

thus, we have

$$a(a-a) = (a+a)(a-a) = 2a(a-a).$$

Dividing by the common factor $a(a-a)$ on both sides, we conclude that

$$1 = 2. \tag{14.1}$$

---

[1] American author of stories of the supernatural, 1842–c1914.
[2] See footnote 1 on p. 1. From *Henry IV*.

(ii) The assertion of the theorem is now easily proved. Let c be the curvature[3] of Earth's surface at any point on it. We subtract 1 from both sides of (14.1) to find that $0 = 1$, or, what is the same,

$$1 = 0. \tag{14.2}$$

Multiplying (14.2) by c, we obtain

$$c = 0$$

and conclude that, since the only surface with zero curvature is a plane, Earth is indeed flat.

Question 2 (requires logic and basic arithmetic). A mathematician meets one of his college friends, whom he has not seen for many years.

"How've you been?" the mathematician asks.

"Pretty good."[4]

"Married?"

"Yep."

"Kids?"

"Three beautiful daughters."

"How old?"

The friend, unimpressed by the other man's terse conversational style, says, "You're the mathematician, you figure it out. Right now the sum of their ages, all positive integers, is 13, and the product of their ages is equal to the number on the house in front of you."

The mathematician looks at the house, thinks for a few seconds and then says, "I need more information."

"My oldest daughter is not allowed to drive a car yet," the friend replies, a little irritated.

Calmly, the mathematician tells his friend the correct ages of his three daughters, adding that his answer is based on the definition of a person's age as the number of birthdays the person has had to date.

How did he arrive at the result?

"Interesting," I said. "Thanks." I folded the paper and put it in my pocket, waiting for JJ to broach, as usual, a subject of his choice. He needed no special invitation.

---

[3] Roughly speaking, curvature is a measure of the amount by which the shape an object differs from being flat. In the case of a circle or a sphere, it is the reciprocal of the radius.

[4] This fellow is obviously not a grammarian.

"Can you guess what arithmetic concept people find most difficult?" he asked me abruptly.

"Percentages?" I offered.

"That's exactly right. Many don't seem to understand what percentages are. They don't know how to compute them or how to interpret them. Percentages frighten some JOEs to death and engender illogical behavior on a colossal scale."

"We mentioned percentages in an earlier conversation, didn't we?"

"We did, but only *en passant*.[5] This topic is worth revisiting."

"For many," I said, "percentages are the main reason why they developed a phobia of mathematics in school."

JJ's eyes narrowed imperceptibly. "What surprises me most is that even some academics have trouble with this simple tool. Listen to the VOTSIT recording of a dialogue between Gan and one of his colleagues, Col, from a university department of arts and social sciences."

GAN: What cost-of-living raise are we getting this year?

COL: It's 8% across the board, half paid from July to December and the other half from next January to June.

GAN: That's not 8% for the year.

COL: Why not? Say that my base salary now is S. Then for the half year from July it becomes S plus 4% of S, which is

$$S + (4/100) \times S = 1.04\,S.$$

And for the half year from January it will be $1.04\,S$ plus 4% of *that*:

$$1.04\,S + (4/100) \times 1.04\,S = 1.0816\,S.$$

As you can see, my base salary at the end of next year will actually increase by 8.16%, which is even better than the official 8%.

GAN (amused but polite): You are wrong. The 4% increase from July is for a half-year; during that period you will be paid

$$0.5\,S + (4/100) \times 0.5\,S = 0.52\,S.$$

From next January, you'll get another 4% for a half-year on top of that, so you'll be paid

$$0.52\,S + (4/100) \times 0.52\,S = 0.5408\,S.$$

Hence, the salary you will receive for the whole of next year is

$$0.52\,S + 0.5408\,S = 1.0608\,S.$$

This means an effective annual raise of 6.08%. Look at the figures in your new contract and compare them to those for the current year. Your base salary will go up by only 6.08%.

---

[5] In passing. (French)

COL: Are you saying that the administrators duped us? They promised an 8% raise for the year!

GAN: Administrators are like politicians: they never tell you the whole truth. And some of them are genuinely innumerate.

"If you want to test a person's intuition about percentages," JJ said, "ask him or her the following simple question. The employees are told by their boss that, due to a shortage of funds, they can choose one of two courses of action for the next two years: (A) they may have a 10% raise in the first year and then a 10% cut in the second, or (B) the other way around. Which should they choose in order to have a higher base salary at the end of the arrangement? Those who don't know how to handle percentages will say, for no reason other than their lack of understanding of what percentages mean, that plus 10% minus 10% will cancel out and leave the original salary unchanged. Some of those who think they have a pretty solid grip on percentages will say that (A) is better because the cut in the second year comes out of a higher salary, therefore leaving them with more, whereas the increase in (B) is applied to a smaller salary, therefore giving them less extra. Only the few who adhere to the principle that mathematics is safest done with a piece of paper and a pen will work out the correct answer, namely, that at the end of the second year all employees, regardless of whether (A) or (B) is used, will have an overall 1% cut in their original salary. This is shown in the following table, computed for a unit of $100."

	At start	First change	Intermediate	Second change	Final
A	100	+10%, or +10	110	−10%, or −11	99
B	100	−10%, or −10	90	+10%, or +9	99

"Of course," JJ added, "practically minded employees will choose plan (A) because it puts $20 more in their pockets than plan (B), the difference coming from what the two plans pay during the first year."

"Ignorance of what percentages are harms an awful lot of individuals," I said. "Particularly some of those who live on debt."

"Living on debt is also a reliable indicator of certain uncomplimentary human characteristics," JJ observed.

"You cannot deny, though, that borrowing may be a necessity at times: it helps people get over temporary financial hardship and improve their standard of living."

"That's fine if they borrow judiciously; if they are aware of the mechanism for servicing their debt and keep their regular payments within affordable limits. Unfortunately, that's not always the case. When it comes to buying, a lot of individuals have much bigger eyes than pragmatism."

"Don't tell me: Gan did a VOTSIT survey about this, too."

"He certainly did. As I explained to you, it's our job to record every aspect of human life. Here are some of the typical answers he got."

GAN: How badly are you in debt?

JOE: Up to my neck. I have five credit cards and reached the limit on all of them. If the card companies increase their interest rates again, I won't be able to meet even the minimum monthly payments.

GAN: How did you come to be in this situation?

JOE: I see other people buying lots of fine things from the stores, and I crave them, too. Since I'm unable to pay their full price up front, I get them on my cards. I have a vague idea that I'll be charged interest on the balance, but I'm bad with figures and cannot compute the interest for myself. When the monthly statement arrives, it horrifies me to see how much the interest actually is.

GAN: What will you do if your payments start falling behind?

JOE: I can always apply for a couple of new cards. And if that doesn't fix it, then I'll contact the helpful people in one of those debt reduction outfits that advertise on TV. They say they can lower my payments. Hey, don't worry about me, buddy: I'll be all right. Life is good."

"I feel sad for these JOEs," I commented. "They don't seem to notice that the helpful people from the so-called debt reduction agencies never say in their TV ads how long they'll be milking them until they pay off their debt."

"Those 'helpful' people are sharks. They get rich by fleecing the unBRITEs of this planet, and they do it with impunity because there is no law to keep their lack of scruples in check. They are in the business of legalized extortion, as are the credit card companies themselves."

"So, you are siding with the debtors?"

JJ shook his head. "No, not really. The debtors who fall on hard times through no fault of their own have my full sympathy. By contrast, as far as I'm concerned, those who are simply financially irresponsible and undisciplined deserve their fate. An unBRITE in this latter category may get out of his predicament in other ways: he may die or successfully rob a bank or start a lucrative drug-pushing business or inherit from a rich uncle or get a higher paid job. But this would be like winning the lottery jackpot. In all probability, sooner or later he will have to declare bankruptcy. And now even that won't make his problem go away: you've seen the new legislation. Anyway, the credit companies don't force people to borrow and spend beyond their means."

"They may not *force* anybody, but they know how to tempt the weak. Have you any idea how many letters they send out to JOEs inviting them to apply for new cards? Think of the number of trees killed to produce the paper for those letters! What a waste!"

JJ nodded agreement. "The card companies don't care about the likes of you and me; they don't make money off us because we settle our bills in full each month; they are targeting the vulnerable, who cannot do that and run into debt, for which they have to pay a lot of interest. Yet therein lies a paradox: sometimes prosperous individuals applying for cards don't get them. Let me show you what happened to Gan when he requested a store card from a national retail chain. He sent in a completed application form under an assumed name and a few days later got a rejection slip. Surprised, he went to see the manager, Man, of the credit account office of that chain. The conversation, recorded on Gan's VOTSIT, went something like this."

GAN: It's about my application. Why was it rejected?

MAN (ignores the question): How did you find my office? Customers are not allowed even to know which town it's in, let alone come here.

GAN (calm and blunt): I tracked down your hiding place because I'm a highly intelligent person, in contrast to your employees, who are unmitigated dummies. The contest was not fair. Now about my application. Why was it rejected? You may want to look at the letter you sent me.

MAN (looks at the letter and types some numbers on his computer): The report from the credit agency that we contacted shows no past activity. You have a blank record.

GAN: And this is bad? To my mind, it would be bad if I had a long record of missed payments and a monumental amount of debt. I'd say a blank record was a positive thing.

MAN: Not when it comes to credit. We have no guarantee that, were you to buy something expensive, you'd be able to pay it back.

GAN: Look at me: do I seem destitute, poorly fed, unhealthy? How do you think I got to be in this top form at my age without having the material means to take care of myself more than comfortably?

MAN: The decision was made by our computer.

GAN: What?? You're using a *machine* to approve or reject credit applications?

MAN: It's very efficient, and its decisions are sound.

GAN: But surely a human can override the machine in special circumstances? Believe me, my circumstances are more special than any you may have come across in your entire life.

MAN: Our machine makes sound decisions. But not to worry: we'll give you a prepayment card. You pay into your account a sum of money and then you can make purchases up to the value you put into it. Then you pay some more—

GAN (loses it a bit): This is adding insult to injury. You're treating me like a potential thief who may default on payments and run away with the merchandise. Want proof that I can pay back what I borrow? Here!

MAN (examines the documents placed by Gan in front of him): What's this?

GAN: The title deeds to my mansion in Hawaii, my penthouse in Manhattan, and my castle in Tuscany.

MAN (returns the documents): Our machine makes sound decisions.

GAN (loses it a bit more and puts a hefty wad of money on the table): And here! A million dollars, which I'm on my way to deposit into my bank account, to add to the other 500 million already there. Now, do you think I'm able to pay back what I borrow?

MAN: Our machine makes sound decisions.

GAN (loses it completely but remains calm): Have you heard of logarithms?

MAN (baffled): What's that got to do with anything?

GAN: I'm the customer. Indulge me.

MAN: Yes, in some math class at college.

GAN: Ah, a college graduate! Splendid! Tell me, if $\ln a = 1$ and $\ln b = 2$, what is $\ln(a+b)$?

MAN (takes a sheet of paper and writes $\ln(a+b) = \ln a + \ln b = 1 + 2 = 3$): There you are.

GAN: Just as expected from an unBRITE with a MICQ no higher than 24: you show complete disregard for the Fifth Mathematical Commandment.

MAN: What does that mean?

GAN: To put it in a language you understand, it means that you are a halfwit, mosquito-brained, sorry excuse for an imbecile. But I may have overestimated your intelligence.

"Too bad Gan didn't win the argument," I said. "I wish he'd prevailed over that unBRITE."

JJ licked his lips. "That's not the whole story. Gan was so maddened by the indignity he had suffered that he took a short vacation from his survey work, acquired a majority of shares in that retail chain, became its CEO, and went again to see Man. This was a much shorter conversation."

GAN: Remember me?

MAN: No, sir.

GAN: You turned down my application for a store card last month.

MAN: Oh, yes.... Our machine always makes sound decisions.

GAN: Not on this occasion. I'm now the CEO of the company, and you, Man, are fired.

"Come on, JJ, Gan didn't do that!" I exclaimed in disbelief. "How did he become a majority shareholder in such a short time?"

"The same way he'd procured the title deeds to those properties and the million dollars: he used what you'd call 'magic'. Naturally, he was tried in our court for gross violation of the noninterference law but escaped the mandatory trip to you-know-where. The judges decided that severe provocation had rendered him temporarily insane. As a consequence, the court showed clemency and reduced his punishment. They put him in charge of recording all the baseball and cricket matches we send to Callisto."

I couldn't help smiling. "You do come up with the most amazing stories, I must say!"

"Don't respond like an ordinary human. Try to keep an open mind about things."

"Well, going back to the business of living on debt," I changed tack, "what would you do if you had the power to diminish or eliminate this dreadful addiction?"

"Two things," JJ explained. "First, I would arrange that, every month for at least a year, no credit card user in the world buy goods in excess of what she can afford to pay that month, and that she pay the balance in full at the end of each monthly grace period. This would earn the credit companies no interest, cut their huge profits, and might make them reconsider some of their policies and practices."

"Fat chance!" I snorted. "They'll find another way to skin their customers. They could, for instance, drastically reduce the interest rates for a short time to make borrowing attractive, then jack them up again. Or some other such unscrupulous scheme. As long as there are fish in the sea, sharks will never die. Besides, even you may not have the technology to do that. Or do you...?"

"We do, but—"

"—you aren't allowed to interfere with the natural development of our species. Too bad. And your second measure?"

"I would revamp the education system to ensure that schoolchildren master the basics about percentages and fractions. Then the credit card companies would have less rip-off fodder." JJ leaned back in his chair. "Now," he said, "it's time for your stories."

I finished the last of the coffee and pushed the empty cup aside. "OK, here we go. Today's two offerings are closely connected with the topic we've been discussing. First one coming up."

A drunk and scruffy-looking man walks into a bank, goes to the first available teller and says to her, "Get your paws in gear and open a savings account for me."

"What was that?" the woman replies politely, taken aback by the man's rudeness.

"Listen to me, you old stiff, and listen good," the man barks out. "Come off your broomstick and open a savings account for me, *now*!"

"Sir, I'm very sorry, but this bank does not condone the use of abusive language on its premises. I'll have to call the manager."

The woman goes into an office at her back and, a few seconds later, emerges in the company of her manager, who steps to the window and asks, "Is there a problem, sir?"

"Not with me, there ain't," the uncouth customer says. "I just won $50 million in the lottery and want to open a stupid savings account in this stupid bank!"

"Very well, sir," the manager says. "And, as I understand it, this stupid woman is refusing to do it for you?"

A wan smile flitted across JJ's face. "Money talks," he said.

"And here's the second story."

A really enterprising reporter sells his soul for an interview with B.L.Z. Bub, which is broadcast live on national TV. After a string of questions regarding the politics of Hell, he asks the devil to identify for the viewers the most despicable deed he had ever committed.

"It's rather difficult to choose one," B.L.Z. Bub says, "for I have done so many. When I saw that people were happy, I visited death and destruction upon them. I brought them famine, epidemics, and wars."

"Is that your worst?" the reporter inquires.

"No, not really. People are very resilient and always recover, however awful the adversity. Then I inserted politics and taxes into their lives."

"Would that be your absolute worst?"

"I don't think so. In time, they learned how to live with one and dodge the other. Then I brought them reformed calculus."

"Is that the lowest of the lowest?

"Close enough, but no."

The reporter doesn't give up. "Come on, your devilship! There must be something so calamitous, so utterly despicable, so horn-withering you did that makes you truly proud above all other deeds!"

B.L.Z. Bub plays with his forked tail and says, "Well, if you really want me to pick an especially evil one, then it'll have to be my invention of the credit card."

"You could claim, I suppose, that the reckless use of credit cards is akin to devil worship." JJ put his hands on the table. "Thanks for the stories," he said. "I've really enjoyed our little chat. If all goes well, we'll talk some more next time." Then, unceremoniously, he sprang to his feet and promptly walked out of the bar.

Notes After the Meeting

Here are the answers to JJ's questions together with some brief personal comments.

Solution to Question 1. What an amusing piece of mathematical garbage! The illegal operation here is cancelation of the factor $a - a$, which is equal to zero. This is rather bad news for the Flat Earth Society.

Not everybody spots the fallacy when they see this 'proof'. It seems that division by zero is alive and well and continues undisturbed in some quarters.

There is a stringent need for all secondary school teachers to impress on their young charges that certain things are simply *not done* in mathematics. Failure to observe such strict taboos could lead to horrific devastation and the untimely death of the universe.

If those to whom you are showing the 'trick' find nothing wrong with division by $a-a$, try changing the argument a little. Start by writing $0 = 0$ (they cannot dispute that), then write $0 \times 1 = 0 \times 2$. It might take them a few seconds to recognize that this is still $0 = 0$, but they should eventually accept it. Once they do, go ahead and chop off the zero on the left and the zero on the right by way of cancelation, and leave them to contemplate the beauty of $1 = 2$. When they start protesting (*if* they start protesting), ask them why they got so agitated when you canceled the factor 0 but were completely unfazed earlier when you had divided both sides by $a-a$, which is just another way of writing 0.

The correct handling is obvious:

$$\begin{aligned} & a(a-a) - 2a(a-a) = 0 \\ \Leftrightarrow \quad & (a-a)(a-2a) = 0 \\ \Leftrightarrow \quad & -a(a-a) = 0, \end{aligned}$$

from which (since $a \neq 0$, as given in the problem)

$$a - a = 0,$$

or $a = a$, a true statement that tells us nothing of interest.

It should be reiterated that this type of manipulation is essential in the solution of equations. Consider the simple example

$$x^3 - x^2 = x^2 + 5x - 6.$$

When all the terms are brought over to the left-hand side and the like terms are sorted out, this becomes

$$x^3 - 2x^2 - 5x + 6 = 0,$$

which, being a cubic equation, would give a lot of grief to the inexperienced solver. In this case, the original form is better because it lends itself to easier factoring:

$$\begin{aligned} & x^3 - x^2 = x^2 + 5x - 6 \\ \Leftrightarrow\ & x^2(x - 1) = (x - 1)(x + 6) \\ \Leftrightarrow\ & x^2(x - 1) - (x - 1)(x + 6) = 0 \\ \Leftrightarrow\ & (x - 1)[x^2 - (x + 6)] = 0 \\ \Leftrightarrow\ & (x - 1)(x^2 - x - 6) = 0, \end{aligned}$$

so

$$x - 1 = 0 \quad \text{or} \quad x^2 - x - 6 = 0.$$

The first possibility yields the root $x = 1$; the second one yields the two roots $x = -2$ and $x = 3$. If the factor $x - 1$ had been wantonly canceled from both sides at an earlier stage, the root $x = 1$ would have been lost.

No one should ever divide by a factor that may be equal to zero.

Some readers may have already noticed that (14.2) could also lead to other 'weird and wonderful' conclusions. For example, we could claim that our physical space is two-dimensional. Let P be the (x, y)-plane in a Cartesian system of coordinates, let Q be an arbitrary plane, and let α be the measure of the angle between P and Q. Multiplying (14.2) by α, we find that $\alpha = 0$, which means that Q is parallel to P. In addition, if d is the distance between P and Q, then, multiplying (14.2) by d, we arrive at $d = 0$, implying that any plane in space—hence, space itself—coincides with the (x, y)-plane.

And so on.

Solution to Question 2. Here is a table listing all possible combinations of the girls' ages that add up to 13, arranged in descending order, together with their corresponding products:

Ages			Product
11	1	1	11
10	2	1	20
9	3	1	27
9	2	2	36
8	4	1	32
8	3	2	48
7	5	1	35
7	4	2	56
7	3	3	63
6	6	1	36
6	5	2	60
6	4	3	72
5	5	3	75
5	4	4	80

This is what the mathematician computed in his mind after he heard the first two clues from his friend, concerning the sum and product of the girls' ages. If the number on the house in front of him had been any one of those in the right-hand column except 36, he would not have needed any more information and would have given the answer right away, because each of those numbers occurs in the table exactly once and, therefore, determines the age distribution uniquely. Since the mathematician asked for more information, it means that the house number he saw must have been the only number in the right-hand column which creates ambiguity, namely 36. This number occurs in the table twice, so it corresponds to two distinct age distributions. Hence, the mathematician needed further clarification, which came in the friend's petulant answer that his oldest daughter was not allowed to drive a car yet. To drive a car in the U.S. one has to be at least 16 (far too young, but that's another issue) and clearly none of the daughters could be of that age. At first glance it would seem that this new bit of information was useless—but it wasn't. The friend's mention of his oldest daughter eliminated the ambiguity because of the two age distributions with a product of 36, that is, $6 \times 6 \times 1 = 36$ and $9 \times 2 \times 2 = 36$, under the definition of age inserted at the end of the problem only the latter allows for an oldest daughter. So, the answer is 9, 2, 2.

The definition of age mentioned above, which implies that two persons are deemed to be of the same age if their birthdays fell within the preceding 12 months, aims to preclude objections from fastidious pure mathematicians. Without it, they may argue that the triple 6, 6, 1 is equally valid because there is always an eldest daughter: even if the two six-year-olds were twins, one of them would have been born marginally ahead of the other. Sometimes, pure mathematicians take great pleasure in spoiling a good riddle.

One final remark: the way I heard the additional clue being given originally was that the oldest daughter had a dog. Since some of the people to whom I gave this problem started asking themselves how old a child had to be to get a dog license, and wanting to prevent such mental excursions into fruitlessness, I decided to change the clue to the car-driving version. Readers should feel free to use their own version when they pass the problem on.

A Word to the Wise

Percentages

☺ **DO**

(i) A percentage is a fraction with denominator 100. Thus, saying that 15% of the cows in a 400-strong herd are red means that the number of red cows in that herd is

$$400 \times \frac{15}{100} = 400 \times 0.15 = 60.$$

Conversely, if it is known that 15% of the cows in a herd are red and the head count of the red cows is 60, we deduce that the number of all the cows in the herd is

$$60 \div \frac{15}{100} = 60 \times \frac{100}{15} = 400.$$

(ii) Great care should be taken when dealing with compound percentages. Suppose that, on the first day of their annual sale, a store reduces the price of a DVD player by 60%, and that, on the second day, it reduces it again by a further 20%. If the price tag now shows $64, how much did the player cost originally?

This type of problem can be worked out in one of two ways.

(a) On the first day, every dollar in the original, unreduced price is discounted by $1 × 0.60 = $0.60 to $0.40. On the second day, this $0.40 is discounted by $0.40 × 0.20 = $0.08 to $0.32. Since the final price is $64, it follows that the original price was $64 ÷ 0.32 = $200.

(b) The final price of $64 represents 80% of the price after the first discount. So, the price after the first discount was $64 ÷ 0.80 = $80. This represents 40% of the original price, which, therefore, was $80 ÷ 0.40 = $200.

☠ **DON'T**

Under no circumstances should the two percentages be added together; they are applied to different quantities. 'Simplifying' the calculation by combining the two reductions of 60% and 20% in the above example into a single one of 80% is plain wishful thinking.

SCAM 15

Modern Art

> *Llamamos arte moderno a esta tragedia grandiosa.*[1]
> Salvador Dalí[2]

> Abstract art: a product of the untalented sold by the unprincipled to the utterly bewildered.
> Al Capp[3]

I met JJ again at a mathematics conference sponsored by The Ohio State University in Columbus, Ohio. Just like the first time, he was sitting alone at a table in a corner of the hotel bar, gazing at me across the room with dark, inscrutable eyes.

As soon as I sat down, he pushed a piece of paper in my direction. "Here," he said laconically. "For your students."

I glanced at the paper and smiled. The text read something like this.

Question 1 (requires basic arithmetic). *Three friends eat at a restaurant and order the same item on the menu. When they finish, they pay $10 each and then leave. As soon as they are through the door, the waiter realizes that he had overcharged them, so he gives $5 to one of the busboys and asks him to run after the customers and give each of them $1 back, which the young man duly does. So, in final analysis, the friends have paid $9 × 3 = $27, which, added to the $2 that the busboy has been left with, makes $29. Where is the remaining $1 from the $30 originally paid by the three friends?*

Question 2 (requires basic algebra). *A computer trick starts by showing you a list of all the integers from 99 to 0, with a letter allocated to each of them, apparently at random (see Table 15.1). You are asked to choose any two-digit positive integer, subtract from it the sum of its two digits, concentrate on the letter marked against your result, and click on a 'magic box' at the side of the list of numbers. As soon as this is done, the screen changes to show the magic box displaying the letter you are looking at. If you are skeptical and want to repeat the experiment, you are invited to click on a 'do it again' button and start afresh. Every single time the magic box produces the correct letter.*
How does this 'mind-reading' trick work?

---

[1] This grandiose tragedy is what we call modern art. (Spanish)
[2] Spanish surrealist painter, sculptor, graphic artist, and designer, 1904–1989.
[3] American cartoonist, 1909–1979.

Table 15.1.

99 U	89 B	79 D	69 E	59 A	49 G	39 G	29 F	19 C	9 D
98 A	88 C	78 Q	68 D	58 P	48 P	38 K	28 V	18 D	8 M
97 T	87 E	77 B	67 I	57 N	47 L	37 J	27 D	17 A	7 F
96 T	86 L	76 J	66 P	56 D	46 Z	36 D	26 Y	16 Q	6 C
95 J	85 T	75 R	65 V	55 C	45 D	35 E	25 L	15 R	5 S
94 F	84 U	74 C	64 H	54 D	44 A	34 R	24 C	14 D	4 K
93 P	83 D	73 U	63 D	53 T	43 H	33 Q	23 A	13 W	3 O
92 L	82 P	72 D	62 R	52 O	42 J	32 B	22 D	12 X	2 V
91 R	81 D	71 L	61 K	51 M	41 S	31 R	21 P	11 Y	1 A
90 W	80 H	70 Q	60 P	50 D	40 F	30 B	20 L	10 G	0 P

"Interesting," I said. "Thanks." I folded the paper and put it in my pocket, waiting for JJ to broach, as usual, a subject of his choice. He needed no special invitation.

"We Jovians," he began abruptly, "have noticed that the progress of your science and technology seems to be in inverse proportion to the quality and depth of your artistic productions. Some of us consider the so-called modern art to be one of the greatest cons perpetrated on the public in recent times."

"If you are about to criticize the lack of logic in the artists' rendition of reality or interpretation of abstract concepts, stop right there: true artistic creation is always energized by a sizable dose of illogic, so it's natural to expect a few weird results."

"'Weird' I buy," JJ said. "What I can't accept is the vulgar, the trashy, and the nauseating. But before you accuse me of not being capable of relating to human modern art because I'm an alien, let me show you a VOTSITed conversation between Gan and two of his students—a math major, Mat, and a creative arts student, Cas."

GAN (to Mat): Have you ever heard atonal music?

MAT: Only once.

GAN: What did you think of it?

MAT (looks uncomfortable): Frankly, I was disappointed. It sounded more like an incoherent jumble of notes than music. My impression was that the composer wanted to vent his anger and frustration on the audience. I consider music to be good when it engages my emotions in a stimulating and pleasing way; when it brings both mind and soul to a state of complete fulfilment, not when it makes me want to attack the piano with an axe. I think that placing such nonsense in the same category of art with the classics borders on sacrilege.

GAN (to Mat): Do you feel the same about modern painting?

MAT: My reaction to that is, well, mixed. I've seen many abstract paintings in museums. In certain cases I like the combination of colors and the patterns, even though I don't have a clue what the whole ensemble is supposed to represent.

Some exhibits, however, look as if they were painted by serial killers. As an arts person, Cas might take a different view.

CAS: An abstract work of art intends to transmit emotions and ideas, not to depict the physical world as we see it. It is therefore appropriate to say that sometimes it may exist only in the mind of the artist.

MAT (with more conviction): But if that's true, then how are we, the general public, going to understand and enjoy it? We operate with our senses and logic; when they cannot successfully process their data, our brain switches to rejection mode.

CAS: Have you ever had the meaning of an abstract painting explained to you by an expert?

MAT: Last year I saw a whole gallery wall painted yellow, with a thin red line drawn diagonally across it. The caption said, "Yellow wall and thin red line joining a point 5 cm from the edge and 10 cm below the ceiling at the top left-hand corner with a point 7 cm from the edge and 12 cm above the floor at the bottom right-hand corner." What explanation do I need for such a 'mural'? It must have taken longer to write the title than to 'paint' it.

CAS: You could interpret it as a homage to the Cartesian plane.

MAT (incredulously): Come on! Then how about a blue rectangle on a white canvas, captioned "Blue rectangle on white background"? What would *that* be: a celebration of quadrilaterals, perhaps? Where is the art? What visual emotion should one feel when looking at it?

CAS: I accept that you didn't feel any emotion. But it doesn't follow that *no one* is moved by it. People see in an abstract painting what they want to see: from nothing at all to many, quite different and quite complex, nuances.

MAT: You mentioned authoritative explanations. I once lingered in front of a painting, trying to understand its message from the legend affixed to its side. After the name of the author and the title, "Sailing on a Somber Soul", there was a text that read something like this: *The black swirl and silver streaks—each rough and laborious, yet purposeful and majestic—reflect the inner torment of the painter, the profound pain seated at the very center of his being. But they also speak of hope: as the peripheral lighter shades suggest, strife is counterpointed by a distinct promise of redemption, a sign that the universe is ultimately merciful.* Where on Earth did that come from? If you ask me, *my* interpretation of what I saw was that the guy had emptied a bunch of tubes of different-colored paints on a canvas and then used the back of his cat to slosh it around. Be honest, Cas: would you pay good money to buy that and hang it in your house?

CAS: I would, if it was signed 'Wassily Kandinsky'.[4]

GAN: This remark is inconsistent with your earlier position, Cas. It implies that you would acquire modern art only if the author was a celebrity, not because you believe in its intrinsic value. Anyway, since you two have divided opinions on the subject, I'd like to issue a challenge. Until our next lecture, find a way of using

---

[4] Russian-born French painter, a pioneer of modern abstract works, 1866–1944.

elements from each other's majoring field to explain and illustrate how you view abstract art. I'm sure you'll come up with something suitable.

"Did they?" I voiced my curiosity. "Mat needed to complete a creative arts project, whereas Cas had to find a mathematical coating for hers. Not an easy task."

"They did. Watch the results."

GAN: Who wants to go first?

CAS: I will. (She taps the computer keys.) My idea was to construct titled ten-digit sequences as an analogy for various periods in the history of art. (Screen comes to life.) The first one:

$$0\ 1\ 2\ 3\ 4\ 5\ 6\ 7\ 8\ 9$$

The Decimal Digit Tree

We easily recognize this as the set of the ten digits arranged in their natural order, much as we recognize the aspects of the real world mirrored in representational art. Next:

$$0\ 2\ 4\ 6\ 8\ 1\ 3\ 5\ 7\ 9$$

Parity

All ten digits are here, but are listed in an unusual order. However, the order is still obvious if we are prepared to make a slight perceptual adjustment: the even figures come first, followed by the odd ones. This brings to mind impressionist art, which, although focused on ordinary subjects, treats them by adding the illusion of movement and unusual visual angles. Next:

$$1\ 2\ 3\ 4\ 5\ 5\ 4\ 3\ 2\ 1$$

Reflective Mood

Here the figures 0, 6, 7, 8, and 9 have been left out, as if unimportant, and the repeated digits are arranged symmetrically. One could argue that this might symbolize modern art, where past traditions are abandoned in favor of experimentation. Next:

$$0\ 2\ 3\ 4\ 6\ 8\ 9\ 5\ 7\ 1$$

A Richness of Divisors

In this sequence, the digits 6 and 8, each with four divisors, are placed in the middle. They are flanked by 4 and 9, each with three divisors. To their left and right are the primes 2, 3 and 5, 7, each with only two divisors. Finally, the end positions are assigned to the 'anomalies' 0 and 1. The original set has been taken apart, analyzed, and put back together in what may at first glance appear to be random order, although there is a theme to it—a characteristic of cubism. Next:

$$0\ 1\ 0\ 3\ 0\ 5\ 0\ 7\ 0\ 9\ 0$$

At Odds

An *eleven*-digit entity! There is structure, order, and symmetry in it, but it defies the prescribed parameters, in a surrealist kind of act that obeys an unconscious and irrational command. Next:

$$1\ 1\ 0\ 0\ 1\ 0\ 1\ 1\ 1\ 0$$

Eight-fourteen

MAT: It's 814 in binary.

CAS: That's right. I would say this sequence can be associated with contemporary art, which often makes use of unconventional materials, displays, computers, algorithms, geometric models, and a whole range of other, quite eccentric, tools and modes of artistic manifestation. And now, for the finale:

$$1\ 0\ 7\ 3\ 7\ 4\ 1\ 8\ 2\ 4$$

Dreaming of Infinity

This seems puzzling and incomprehensible. It does not contain all the digits—5, 6, and 9 are missing, 1, 4, and 7 occur twice—and there is no visible symmetry or order in its structure. You might say that it's merely an arbitrary, meaningless collection, a nihilistic act of rebellion and anarchy against the established norms, just as most of abstract expressionist art appears to the public at large. But not to the connoisseurs. (Turns to Mat.) Okay, Ramanujan,[5] are you a connoisseur or not?

MAT (strains his eyes and concentrates for a few seconds): This is 2 to the power of 30.

CAS (impressed): How did you know?

MAT: An educated guess. The number ends in 824, which makes it a multiple of 8. This suggests a power of 2. Since $2^{10} = 1024$, 2^{20} has seven digits, begins with a 1 and ends in a 6, and 2^{30} has ten digits, also begins with a 1 and ends in a 4, I took my chances.

CAS: And you were right. Now, I hope, you will be more charitable toward abstract art for, as you can see, sometimes you do find meaning in it if your senses and feelings are tuned to the right frequency.

GAN: Thank you, Cas. You did a very good project and proved your point most elegantly. So, Mat, what have *you* got for us?

---

[5] Srinivasa Ramanujan, Indian mathematician, 1887–1920. Among other professional skills, he was reputed to have an unusual ability to recognize obscure properties of numbers.

MAT: The only creative art form I could handle—if that's the right word—is literature. We know about the theater of the absurd and its message that life has no meaning, but don't worry, I didn't write a play. Instead, I tried my hand at abstract poetry, where the sound of words takes precedence over their significance. Listen to it. (Smiles and starts reciting.)

> Prime Rhyme
>
> In compliance, first comes two,
>   Spooky
>   Rookie
> Dressed in blue.
>
> Later on, discover three,
>   Gory
>   Story
> Told with glee.
>
> Full of thunder, next jumps five,
>   Spunky
>   Monkey
> All alive.
>
> Not enough? So look at seven,
>   Handsome
>   Ransom
> Paid in Heaven.
>
> Now take stock, and from the four,
>   Tender
>   Splendor
> Get lots more.

GAN (to Cas): Care to critique?

CAS: Hmm.... The technique is solid, the rhyming sounds have variety and are created by healthy adjective–noun pairs, and, surprisingly, there is a theme going with the verse: Mat clearly talks about the first four prime numbers, that is, 2, 3, 5, and 7. Then, at the end, he springs a puzzle on the listener. It appears that combining these numbers in some way we can obtain more primes. My brain is not a fast calculator, so I need to ask: is 2357 a prime?

MAT: It certainly is. And there are seven more prime permutations of the four digits: 2753, 3257, 3527, 5237, 5273, 7253, and 7523.

CAS: Eight in all! That's spectacular! Well done!

MAT: Thanks for the endorsement, Cas. (Smiles again.) But I'm afraid there are *nine* combinations.

CAS: Nine?! Wait a minute.... (Thinks briefly, moving her lips.) You are right: we also have the sum $2 + 3 + 5 + 7 = 17$, which is another prime. Sheer brilliance, Mat! Will you marry me?

"Cas and Mat are smart, numerate, well-educated young persons with a good sense of humor," I said approvingly. "Did Gan measure their MICQ?"

"Over 60."

"I'm not surprised." My eyes leveled on JJ. "Earlier, you said something rather harsh about our current art scene. Why?"

"I and my fellow Jovians don't believe that all its creations are genuine. According to our evidence, some of your 'artists', diagnosed by the VOTSIT with a MICQ no higher than 39, are deceiving the public."

"Perhaps. But it's up to the individuals to decide whether they're being taken for a ride. If they feel that contemporary art is straying from their values, they're free to shun it."

"It's not so simple. According to another one of Gan's surveys, quite a few JOEs share our views but, out of snobbery, or to avoid being dubbed philistines, they pretend to march in step with the cultural avant-garde and affect an understanding and appreciation of artistic modernism. The nakedness of Andersen's emperor[6] continues to remain unacknowledged."

"Do you think mathematics has anything to do with art?"

"You mean, apart from underlying the construction of harmonies, the interplay of colors, and the purity of architectural forms? Of course it does! Gan and his students have had many discussions on this topic. Here he is again with Mat and Cas.

MAT: Dr. Gan, can you detect any connection between art and mathematics?

GAN: I think that certain mathematical elements do find a place in art. For example, experts say that the golden ratio can be identified as a structural component in the works of some architects and musicians.

CAS: What is the golden ratio?

GAN: Consider two distinct positive numbers a and b such that a is greater than b and
$$\frac{a+b}{a} = \frac{a}{b}.$$

The number φ defined by either side of the above equality is called the golden ratio, or golden section. Writing the equality as $1 + 1/\varphi = \varphi$, we easily see that $\varphi = \frac{1}{2}(1 + \sqrt{5})$, which is an irrational number, approximately equal to 1.618.

MAT: And this number has aesthetic value?

GAN: Apparently so. To take a simple case, it is claimed that a golden rectangle—that is, a rectangle with the ratio of its longer side to the shorter one equal to φ—is very pleasing to the eye. Apart from mathematicians and artists, φ is also of great interest to historians, psychologists, and biologists.

MAT: Have any other mathematical objects been observed to influence artistic work?

---

[6] The central character of the fairy tale "The Emperor's New Clothes" by Hans Christian Andersen, Danish author and poet, 1805–1875.

GAN: The pundits are excited by the Fibonacci[7] sequence.
CAS: What is that?
GAN: It's an infinite sequence that starts with 0 and 1 and where each term after that is equal to the sum of the preceding two:

$$0, 1, 1, 2, 3, 5, 8, 13, 21, 34, 55, 89, 144, 233, 377, 610, \ldots.$$

This is written in compact form as

$$F_0 = 0, \quad F_1 = 1, \quad F_n = F_{n-1} + F_{n-2}, \quad n = 2, 3, \ldots.$$

Strangely enough, the Fibonacci numbers are related to the golden ratio. The connection is given by the formula

$$F_n = \frac{1}{\sqrt{5}} [\varphi^n - (1 - \varphi)^n].$$

The sequence appears to have been used in the work of some architects, poets, musicians, quilt designers, and movie makers.

MAT: What about fractals? I've seen posters with them, and they look quite artistic.

GAN: You're right: they do. In fact, there is such a thing as fractal art.

CAS: Are fractals mathematical objects, too?

GAN: Roughly speaking, a fractal is a self-similar geometric shape; in other words, it consists of parts that are smaller-scale copies of the whole. A fractal curve is sometimes constructed recursively. (Works the computer keys.) Here is a simple example, called the Koch[8] snowflake.

---

[7] Leonardo di Pisa, known as Fibonacci, Italian mathematician, c1170–c1250.
[8] Niels Fabian Helge von Koch, Swedish mathematician, 1870–1924.

The recursive process, shown with its first six iterations, is self-explanatory. Notice that it's practically impossible to see the difference between the curves in iterations 5 and 6 if you discount the 'heaviness' of the line, which is due to the richer microstructure of the latter. This difference becomes apparent only if you magnify the figures.

MAT: I have a feeling that, as the iterations continue, the length of this curve tends to infinity.

GAN: You are absolutely correct.

CAS: But the area it encloses remains finite!

GAN: You, too, are correct. Well observed. An intriguing paradox, isn't it? A simple closed curve of infinite length surrounding a region of finite area. Fractals have strange properties. When colors are introduced, they can make very pretty and unusual sights. No wonder some modern artists' work includes the production of fractal pictures and even fractal music. Needless to say, this is all done by computer, not by hand. The artist, however, has control over the iterated function and some other parameters.

MAT: Dr. Gan, we get the idea how mathematics influences art. Is there any 'traffic' in the other direction?

GAN: Yes, although it's more vague and underdeveloped. To put it succinctly, the intrinsic order present in the structure of musical composition and the relationships between proportion, shape, and color found in painting are not unlike the properties of the building blocks of mathematical proofs. Add to them their direct influence on human perception, and you have a fascinating and rewarding area of exploration for mathematicians.

"But this is the view from *our* side," I commented. "Do artists feel the same?"

"I think that, while interested in what science can provide for them in the way of novel means of expression, they don't really care whether or not we study, quantify, and interpret the results of their labor. Many of them don't believe that mathematics by itself is capable of creating art because it cannot generate human thought and emotion."

"Are they right?"

"Most certainly not. All sentient beings are part of the universe; therefore, everything they conceive, execute, and feel can, in principle, be analyzed, explained, and synthesized by mathematics. We just haven't yet discovered how; in fact, we may never do so before the universe comes to an end. The potential, however, is always there."

I leaned slightly toward the table. "Mat wrote his quirky verse using everyday language. But what if we crossed over into the abstract world of mathematical symbolism? How would a story sound in that type of setting?"

"I've got a specimen for you," JJ said. "Gan amused himself one day and came up with something like this."

A long time ago, in a normed space far, far away (with respect to the metric induced by the norm), there lived a subset of continuous functions of x defined on the same closed interval. The functions were happy in each other's company and coexisted peacefully, intersecting their graphs at times and occasionally getting composed with one another to produce even more functions.

Life in the subset would have seemed sheer bliss had it not been punctuated by debilitating moments when the differentiation operator appeared out of nowhere to wreak havoc on that little corner of paradise. At such moments, the functions would recoil in fear and scurry around looking for cover, for they knew that, once the operator got attached to them, they would lose smoothness and be forced to undergo an indelible change of personality. All, that is, except one called e-to-the-x, who always held her[9] ground and shouted at the marauding operator with supreme arrogance, "I'm e-to-the-x, so you can't touch me!"

And indeed, at the end of every raid e-to-the-x was still standing there, whole, proud, and defiant, structurally unaltered.

"I hate that boastful simpleton," function after function took to muttering in compact recesses of the subset, although, grudgingly, they all admired and envied her untouchable quality. "I wish that differential monster had more bite, to teach her a lesson. May she end up in the null space of a linear transformation!"

In the material universe, generally speaking, there is no proof that curses have a real effect on their subject. However, from time to time, whether by sheer coincidence or due to some specific but unfathomable causes, they do seem to work. The same is apparently true in normed spaces. And so it came to pass that one fateful day, when everything in the subset was sweet and quiet, the old enemy burst forth and went on the rampage again.

While everyone else panicked, e-to-the-x looked around and shrugged with boredom. "Differential operators!" yelled a piecewise smooth function, trying forlornly to hide her points of nondifferentiability. "Two of them! Run for your life!"

But e-to-the-x did nothing of the sort. Examining the scene, she coolly took in the mayhem and waited for it to subside.

There was sine-of-two-x, undulating sluggishly about. One of the operators rushed at her: wham! and suddenly sine-of-two-x shifted by a half-period and doubled her amplitude, becoming two-cos-of-two-x.

Away from the origin, square-root-of-x was trying to take cover when that same operator reached her: wham! and she immediately had the origin removed from her domain, which triggered her instant expulsion from the subset.

While the first operator produced his customary share of devastation maiming functions left, right, and center, the second one seemed to clear everything in his path. Slowly, ineluctably, he came toward e-to-the-x and stopped in front of her.

"I'm e-to-the-x, so you can't touch me!" the well-rehearsed battle cry went out.

The operator smiled with faked kindness, as operators do when they hear something amusing, brandished his unforgiving incremental ratio, and said, "Delighted to meet you, my dear. I'm d-by-d-y."

---

[9] To make things somewhat more intuitive, in this little story functions and operators have been assigned feminine and masculine gender, respectively, in keeping with their Latin origin.

Wham! and e-to-the-x was no more.

Which goes to show that when you are a function, you must always check a derivative's operational variable if you want to avoid obliteration.

"Small wonder some people think that mathematicians are a bit peculiar!" I laughed.

"Only those who buy modern art indiscriminately think so. Anyway, to make my story fully logical, I should've specified that x was independent of y. What an insufferably precise lingo we have!"

"Do you see mathematicians actually *write* and *publish* this sort of thing?"

"Good grief, no! If they did, they'd very quickly run out of friends. Even worse, if they have an uncompromising dean, they might find themselves separating variables for food on street corners." JJ leaned back in his chair. "Now," he said, "it's time for your stories."

I finished the last of the coffee and pushed the empty cup aside. "OK, here we go. Today I've got three. The first one is a two-liner."

Question: what is the similarity between a chicken?
Answer: it has both legs parallel, especially the left one.

"Surrealist humor!" JJ exclaimed.
"Not something you hear too often. And now the second one."

In the honeymoon suite of the old hotel, deep in the Transylvanian mountains, the young bride woke up in the middle of the night with a strange foreboding. The room was quiet. At her side, the groom breathed softly, a sound of tranquil contentment. Although the darkness felt velvety and warm, it exuded an unease that made her wide-open eyes search for reassuring familiar shapes. She cringed slightly when none was found.

As her mind raced back to the stories told by the wedding guests earlier in the evening, she heard a rumble in the distance, like the signature of a faraway storm, yet different. At first it seemed to hang in the air, confused and remote. Then, with a suddenness that took her breath away, the noise intensified beyond endurance, rushing at her in a mad, lugubrious howl, a colossal juggernaut of doom that cascaded thunderously about her ears, crushing her thoughts and chasing their distorted fragments out of her mind.

Panicked, she reached for the phone on the nightstand. The line was dead. She was about to get out of the bed and go to the door when the four-poster started shaking violently. Her teeth chattered with fright. A sense of evil invaded the room, overwhelming, irresistible, all-powerful. Those stories and legends.... How she had laughed, calling them stuff for scaring little children! The old people seated nearby had nodded their heads and drunk the wine from the tall, slender glasses, making no comment. Could they have been right? Could they?

Baffled that her groom was oblivious to the commotion, the bride wanted to wake him up. But she had no time. A blinding red flash split the night, hurting her eyes, and a terrifying, unearthly cry bounced loudly from wall to wall. Shocked and frightened to the core, she matched that cry with one of her own, born of anguish and incomprehension.

It all lasted but a moment. When the darkness returned and the eerie wailing died down, she pressed the switch on the headboard and turned on the lamp. Then she froze: the groom was no longer there! Her cheeks drenched in tears, she began to search the room. She looked carefully everywhere but found no trace of him. All his things had vanished, too, as if they had never been there. Exhausted, she threw herself across the bed and wept, her body shaken by pitiful convulsions.

A while later, she wiped her face and searched the room again, praying for a sign that might let her piece the puzzle together, a clue as to what had happened. And this time, when her arm panned under the bed, she touched something solid. With trembling fingers, she brought it out. It was a shoe. Would this lonely shoe, she wondered, hold the answer she was so desperate to have? Would it help her unravel the horror and the mystery?

The shoe felt hot to the touch. She held it gently and examined it with great thoroughness, but found no sign. Finally, she turned it over. And there, on the smoldering sole, she saw the inscription, the fiery letters that would remain burnt into her mind and heart for the rest of her life: *Made in Italy.*

"A feast of the absurd!" JJ enthused.

"Last story reporting for duty. Same subject, different angle."

After his divorce becomes final, Jim asks his friend Tom to arrange a blind date for him, which Tom obligingly does. The day after, Jim telephones Tom and gives him an earful.

"Are you crazy or did you just want to have a good laugh at my expense? Why did you pick *this* woman? Her eyes were squint and of different colors, her nose was long, thin, and bent like a witch's, her hair was growing on her cheeks instead of her head, her chest was flat, and her legs had the same diameter all the way from her ankles up. You call this a date?"

Tom listens patiently, then says, "I don't know why you're complaining. You always boast how much you appreciate and understand modern painting. Well, you can't have it both ways: either you like Picasso,[10] or you don't."

"I think Tom's got a point there, one that surely escapes a politically correct person who reads the physical description of Jim's blind date." JJ put his hands on the table. "Thanks for the stories," he said. "I've really enjoyed our little chat. If all goes well, we'll talk some more next time." Then, unceremoniously, he sprang to his feet and promptly walked out of the bar.

Notes After the Meeting

Here are the answers to JJ's questions together with some brief personal comments.

---

[10] Pablo Ruiz y Picasso: Spanish painter and sculptor, cofounder of cubism, 1881–1973.

Solution to Question 1. There is no $1 left over. The apparent discrepancy is caused by the incorrect manipulation of the numbers involved. The sums are properly tallied up like this:

(i) in the original transaction,

$$\$10 \times 3 = \$30 \ (paid) = \$25 \ (in \ the \ hands \ of \ the \ waiter)$$
$$+ \$5 \ (given \ to \ the \ busboy);$$

(ii) after the refund,

$$\$9 \times 3 = \$27 \ (paid) = \$25 \ (in \ the \ hands \ of \ the \ waiter)$$
$$+ \$2 \ (left \ with \ the \ busboy).$$

Friendly advice for the reader: if you think you have been overcharged in a restaurant, then, first, make sure of the facts; second, query the bill; third, do not leave a tip; and fourth, never eat there again unless you don't mind a server with a bad attitude.

Hell hath no fury like a waiter untipped.

Solution to Question 2. This is trivial algebra combined with a neat sleight-of-hand. Let $\overline{ab}$ be any two-digit positive integer. In terms of the digits a and b, the number is written as $10a + b$. Then the operation to be performed on it is

$$(10a + b) - (a + b) = 9a. \tag{15.1}$$

Since a is restricted to one of the integral values 1, 2, 3, ..., 9, it follows that the only possible results in (15.1) are 9, 18, 27, ..., 81. If you look carefully at the table on the screen, you'll notice that these numbers have against them one and the same letter, which is the letter the magic box will show you. For the example in Table 15.1, this letter is D.

A click on the 'do it again' button takes you back to the first screen, where the letters in the list have now been rearranged so that the multiples of 9 are all marked again by a single letter, but different from the one before. Assuming you have done sum (15.1) correctly, it is this new letter you'll see in the magic box. The letter reshuffle is intended to make it more difficult for the player to work out the algorithm.

Persons with less mathematical experience who don't believe in Santa Claus normally try the trick several times until they spot that the only results they are getting from the required bit of arithmetic are multiples of 9. This observation practically unlocks the door to the solution.

A mathematician follows the instructions once, mentally derives formula (15.1), checks the letter against the multiples of 9 in the table, and then goes off in search of a new challenge.

A few of the engineers to whom I showed this trick found the solution in two steps. First, they conducted a lot of experiments, repeating the trick over and over and then concluding that it worked correctly every time. Second, they resorted to the sledgehammer method (or the method of proof by exhaustion) to work out the algorithm: they wrote a code that took all the numbers from 10 to 99, worked out the necessary arithmetic for each of them, realized that the answers they got were always multiples of 9, and then spotted that all these numbers had the same letter against them in every list. Time-consuming, but correct. Who says that, when solving a mathematical problem, one has to follow the reasoning geodesic?[11]

A Word to the Wise

Exponents

 DO

The following are legitimate operations with exponents:

$$a^p a^q = a^{p+q},$$

$$\frac{a^p}{a^q} = a^{p-q},$$

$$(a^p)^q = a^{pq},$$

$$(ab)^p = a^p b^p.$$

Here a and b are any positive numbers and p and q are any real numbers.

 DON'T

The power function and exponential function *do not* behave linearly; therefore, do not write

 $(x+y)^a = x^a + y^a,$

 $a^{x+y} = a^x + a^y.$

---

[11] In mathematics, a geodesic is the shortest path between two points in a curved space.

SCAM 16

Averages and Buying Cars

> *The salesman knows nothing of what he is selling save that he is charging a great deal too much for it.*
> Oscar Wilde[1]
>
> *Everyone lives by selling something.*
> Robert Louis Stevenson[2]

I met JJ again at a mathematics conference sponsored by Princeton University in Princeton, New Jersey. Just like the first time, he was sitting alone at a table in a corner of the hotel bar, gazing at me across the room with dark, inscrutable eyes.

As soon as I sat down, he pushed a piece of paper in my direction. "Here," he said laconically. "For your students."

I glanced at the paper and smiled. The text read something like this.

Question 1 (requires basic arithmetic or algebra). *The average age of the students in a class is 13 years. When a new, fourteen-year-old student joins the class, the average age jumps up to 13.05 years. How many students are now in the class?*

Question 2 (requires calculus). *What, if anything, is wrong with the following assertion and proof?*

Theorem. *Anyone can own a brand new Porsche.*

Proof. (i) First, we demonstrate that 0 is the only number. Consider the sum (called an infinite series, because it has infinitely many terms)

$$\sum_{n=1}^{\infty}(-1)^{n-1} = 1 - 1 + 1 - 1 + 1 - 1 + \cdots.$$

Grouping the terms in two different ways, we have

$$(1-1) + (1-1) + (1-1) + \cdots = 0 + 0 + 0 + \cdots = 0,$$
$$1 - (1-1) - (1-1) - \cdots = 1 - 0 - 0 - \cdots = 1;$$

therefore,

$$1 = 0. \tag{16.1}$$

---

[1] Anglo-Irish writer, 1854–1900.
[2] Scottish writer, 1850–1894.

Let x be any number. Multiplying both sides of (16.1) by x, we find that

$$x = 0. \tag{16.2}$$

(ii) Suppose now that a customer turns up at a Porsche dealership and finds out that the price of a brand new car is $\$x$. Formula (16.2) tells him that, in fact, he can have the car for nothing, so he buys it on the spot and takes it home. Since anyone can do that, the theorem is proved.

"Interesting," I said. "Thanks." I folded the paper and put it in my pocket, waiting for JJ to broach, as usual, a subject of his choice. He needed no special invitation.

"I've heard people say," he began abruptly, "that, on average, men date more than women. You know that this is not true."

"It certainly isn't. But it's hard to change the general public's belief. Men usually feel macho and like to boast of their prowess in getting dates, whereas women, more inclined to privacy, tend to be quieter on the subject—hence the fallacy. Have you thought of a simple way to set the record straight?"

"The simplest is to use what you just mentioned: a record. For a credible argument, we need to look at a constant and unbiased population of men and women in equal numbers, over a given interval of time; to make the argument easy to grasp, let's assume that the population consists of 6 women, A, B, C, D, E, and F, and 6 men, M, N, P, Q, R, and S, and that, during one month, we gathered the following information on their dating:

	A	B	C	D	E	F
M	0	1	2	0	1	2
N	5	1	3	4	1	2
P	0	1	0	2	2	3
Q	0	0	0	0	4	0
R	0	1	0	1	0	0
S	0	0	0	0	0	0

This means that, for example, M had 1 date with B, 2 dates with C, 1 date with E, and 2 dates with F, C had 2 dates with M and 3 dates with N, and so on. Adding the entries in each row, we obtain the number of dates had by each man:

M	N	P	Q	R	S
6	16	8	4	2	0

Adding the entries in each column, we obtain the number of dates had by each woman:

A	B	C	D	E	F
5	4	5	7	8	7

It is, therefore, immediately clear that the total number of the men's dates and that of the women's dates are exactly the same, 36, which gives the same average of 6 dates per person. The conclusion is independent of the size and entries of the table."

"It is also interesting to observe," I added, "that if the numbers of men and women in the population are unequal, then the averages for the two sexes may differ, but the total number of dates will still be the same for each sex. Individually, however, things are always uneven. And in the case proposed by you we can also draw some other, nonmathematical conclusions: N seems to be very much a lady's man, A is a one-man woman (but, apparently, made the wrong choice), whereas S needs to clean up his act if he wants to get any date at all."

"Useful thing, averages. When properly manipulated, they can be very profitable. Gan bought a car the other day and did exactly that: he put the definition of average to good use."

"New car or pre-owned?"

JJ gave me what I took to be a reproachful look. *"Et tu, Brute?"*[3] Such a ridiculous euphemism! A secondhand car by any other name is still a secondhand car. Calling it 'pre-owned' does not shave a single day off its age or improve its mechanical quality. People favoring this label probably believe that political correctness should be extended to vehicles as well. No, Gan bought a new motor." He paused, drumming his fingers on the table, then said, "Why do car salesmen start acting like predators the moment you step on their lot? This unBRITE behavior, fueled by a twisted operational logic, is what prevents their MICQ from rising higher than 19."

"They work on commission. When their 'radar' flags up a potential customer, they rush to turn him into a real one. Did Gan use the VOTSIT in his negotiations?"

"Certainly not. It would've been unethical. We are not allowed to elicit information from persons against their will and then use it to impede them in their jobs or to get an unfair advantage for ourselves. Besides, Gan looked upon this experience as a challenge. He used the device only for recording, and then briefly on full function at the end, to see what kind of a deal he'd done."

"Was he pleased with the result?"

"Judge for yourself," JJ said. "Here is the gist of Gan's conversation with the salesman, Sal, and his manager, Man."

SAL: Hi! Nice car, isn't it? Good mileage per gallon as well.
GAN: Yes, it's a nice model.
SAL: A lot of people buy it. Will you tell me your name?

---

[3] *You too, Brutus?* (Latin) A line from Shakespeare's play "Julius Caesar", used to describe betrayal.

Gaius Julius Caesar, Roman military and political leader, 100–44 BC.

Marcus Junius Brutus, Roman senator and Caesar's close friend, 85–42 BC. He was a member of the assassination plot against Caesar.

GAN (with Ganymedean bluntness): It's Sir. How much are you asking for this car?

SAL (leans to the car's near window and reads the figure on the paper stuck there): 29,995 dollars, sir.

GAN: To all intents and purposes, that's 30,000 bucks.

SAL: No, sir, it's only 29,995.

GAN (looks sternly at Sal): To all intents and purposes, it's 30,000.

SAL (probably remembers that the customer is always right): Yes, sir, it's 5 dollars short of 30,000.

GAN: When I asked how much it costs, I didn't want you to tell me the figure written on the ticket. I could've read that myself. I meant the price you *realistically* expect for this car.

SAL: As I said, sir, 29,995—

GAN: Young man, we don't seem to be speaking the same language. We aren't communicating here. You think I fell to Earth from another planet? Who ever pays the MSRP printed on the original ticket?

SAL (feigns understanding): Ah, sir, you are actually interested in *buying* the car. I thought you were just browsing. In this case, let's go to my office and have a chat about it.

GAN (a few seconds later, sitting across from Sal in a little glass-walled cubicle): So what's the real number?

SAL: For customers like yourself, we can shave off 1,000 dollars. It's a very good discount.

GAN: You can, can you? Listen, you give this discount to everybody—I saw your TV ad. What I want to know is how much *further* you are prepared to discount the price.

SAL (bends over the desk and whispers): For truly *good* customers who know their way around cars, we can take off another 500 dollars. But, you understand, this is strictly unofficial. The manufacturer doesn't want us to do it.

GAN: Fine, we aren't going to tell the manufacturer. So now we have a starting point of 28,500.

SAL (with a jolt): *Starting* point, sir?

GAN (shows Sal a printed sheet): Indeed. Here is a copy of this dealership's web page, where it says that there is a charge of 600 for bringing the car over from the regional distributor.

SAL (defensively): All cars carry that charge, I'm afraid...

GAN: Don't worry, I'm not asking you to waive it. But look at this: what is the 500 added on the next line for? What's this 'improved trim kit'?

SAL: It includes a special foot mat for the driver and one or two other little luxuries.

GAN: Were they fitted by the manufacturer as standard?

SAL: Not for this model. We put them in ourselves.

GAN: I don't need them.

SAL: But they are already installed in the car.

GAN: I don't care. Take them out.

SAL: I can't. It would mean trashing them. They've been used.

GAN: Then order me another car from the distributor and make sure you don't add anything to it unless instructed by myself, the *buyer*.

SAL (looks distraught): This will take at least 6 weeks! We'll be well into the next sales quarter. Can we not come to some sort of compromise and conclude the sale today?

GAN: We could. But first you must remove the unwanted kit or agree not to charge me for it.

SAL: That's beyond my authority, sir. I need to call the manager.

MAN (arrives a few seconds later): What seems to be the problem?

SAL (explains Gan's demand): We are looking for a way forward, Mr. Man.

GAN (to Man): Before we resolve the kit issue, I want to ask about the other additional charge: 500 for paperwork. What's that for?

MAN: There are forms to be filled by our secretaries in connection with the purchase of any vehicle. We need to pay their salaries.

GAN: *You* need to pay their salaries, not me. I want that charge taken off as well.

MAN: But I need to pay their salaries every month—

GAN: Tell me: do these salaries vary from month to month? Are they higher in a month when the sales soar and lower in a month when the sales dip? If in one particular month you do not sell any cars, do your secretaries get anything at all?

MAN: They get paid the same every month.

GAN: So their pay is not proportional to the volume of sales. It's not linked to the number of 'paperwork fees' you are charging the buyers.

MAN: Our secretaries' pay is averaged out over the year.

GAN: Good, I'm glad to see that you understand averages. But the charge is exorbitant. How much does a secretary make? Let's say that you are generous and pay them $20 an hour. A fee of $500 means that a secretary spends 25 hours filling in the forms for every car sold. Are your secretaries complete dumbbells who cannot type faster than a word per minute?

MAN: We also pay property taxes for the premises, heating bills—

GAN: Then you are guilty of misrepresentation when you are calling this a 'paperwork fee'. It seems to me that the fee is used as a pretext to push prices up, and I'm not prepared to pay it.

MAN (sits down): Sir, it is clear to me that you are one of those valued customers who does his homework before coming in to buy a car. Let's not argue about individual items—let's just come up with an overall price that pleases you and allows my salesman to earn enough commission to be able to afford a loaf of bread for his hungry kids tonight.

GAN (amused): Very well. Let's put our cards on the table. I know that you cannot sell below cost, for if you did, then both you and Sal would lose your jobs. So the question is what margin of profit is reasonable for you and acceptable to me. Well, since the exact cost of production and delivery is never disclosed by the manufacturer, the best guidance we can get is from the price paid by other buyers of this model in the last, say, three months.

MAN: And how do you propose we find out what that is?

GAN: There are very useful and credible web sites where such information is readily available. Have a look at this.

MAN (examines another piece of paper produced by Gan): 27,000! But this is daylight robbery! You want me and Sal to become destitute, lose our homes and families, and sleep in cardboard boxes under a bridge? We cannot sell you the car at this price!

GAN: Well, if you put it like that, then I don't have any option but to take my business to another dealership in town.

MAN (wringing his hands): Sir, you are driving a fierce bargain. Please let me think for a second.... All right, all right: against my better judgement, I'll give you the car for 27,000, but you must promise never to tell a soul about it. If word gets around, I'll go bankrupt. Is this a great deal for you or what?

GAN (points to the paper): Not so fast, Mr. Man. See this word here, in front of 'price: $27,000'? The word is 'average'. 27,000 is the *average* price paid for this car in the last three months. You showed me earlier that you know what *average* means. To avoid any misunderstanding, let's clarify this once more, in the context of car sale prices. Can you give me an example?

MAN (eagerly): Yes, sir. Suppose, for simplicity, that we have a total of six buyers. If each of them paid exactly 27,000, the average price would obviously be 27,000. But a same-price situation is highly unlikely since the six buyers got their cars from different dealerships around the country. This means that some buyers paid more than 27,000 for it—for example, 27,100, 27,300, and 27,500—whereas others paid less—for example, 26,900, 26,700, and 26,500. The average, equal to the sum of all these prices divided by their number, is, of course, 27,000.

GAN: Okay, so we are clear on the definition of average. You are also right about it being very likely that some buyers would have paid more and some less than the average price. I must tell you that I'm the kind of buyer who pays less.

MAN (pretending to be shocked): I'm shocked! A man of your intellect behaving so illogically!

GAN: On the contrary, it's sound logic that informs my decision. In your example, you picked the six numbers clustered symmetrically and very close to their average. Now suppose that one of the buyers went wobbly at the knees and paid the full manufacturer recommended price of 30,000, while the other five paid 26,400 each. The average is once again 27,000. However, were I to pay 27,000, I would make a bad deal because I would be giving you 600 more than the market-realistic price of 26,400. This is why I will offer below the average.

MAN: But, sir, the truth could be quite the opposite, couldn't it? One buyer may have paid only 24,000 and the other five paid 27,600 each. Then paying 27,000 would get you a great deal.

GAN: Mr. Man, your argument is self-defeating. If someone paid 24,000 for this car, I would logically conclude that, since no car dealer sells his wares at a loss, the price of 24,000 contained an acceptable margin of profit and my offer would be pitched at that level. In your own interest, I advise you to stay on my side of logic.

MAN (tries hard but fails miserably to make himself cry): Sir, I don't believe this! After all the concessions I've made to you, you're now throwing my kindness in my face? Have you no compassion at all? Are you not human like us? In all my years in this business, I've never had a customer wanting the skin off my back. Let me bring you a knife so you can do me a favor and cut my throat.

GAN (undaunted): My final offer is 26,500. Take it or leave it.

MAN: Perhaps you could go as far as 26,750 just so at least I don't subsidize the car out of my own pocket?

GAN (gets up and gathers his papers): It's not negotiable.

MAN (dabs at the corners of his eyes): Very well, sir. You win. You can have the car. But I won't forget you in a hurry. You've driven me into the ground. I feel as if life is not worth living any more.

GAN (presses the truth stimulator button on the VOTSIT in his briefcase): Histrionics apart, Mr. Man, how far down were you prepared to go on the price?

MAN: 26,300.

GAN (in Ganymedean): $&@%@$!

MAN (confused): What?

GAN (switches off the device): Nothing. I was saying that, since I don't see you pulling your hair and writhing on the floor in pain, you've still come out ahead in this little charade.

MAN (more composed now): I have my pride, sir. I'll wait until I get home and hold a pillow over my dear wife's face before I do that.

"A car salesman with a sense of humor!" JJ marveled. "Who would've thought such a species existed?"

"So in the end your colleague paid 200 more than Man's bottom price."

"Yup. But, operating within his set parameters, Gan was pleased with the result. Had he made a bid below 26,300, Man would have answered an emphatic 'no' and walked away. This would've left Gan in a weak position if he still wanted to buy the car, which he did. As the Chinese put it, you cannot beat the local snake." JJ leaned back in his chair. "Now," he said, "it's time for your stories."

I finished the last of the coffee and pushed the empty cup aside. "OK, here we go. Today I've got three. First one coming up."

A mathematician and a computer scientist are traveling together on a plane from New York to Los Angeles, returning from a conference. It is a night flight, and after takeoff, the mathematician, very tired, tilts his seat back and wants to have a nap.

The computer scientist, who, against advice, had too much coffee after dinner, is restless and craves company. He tugs at the mathematician's sleeve and says, "Let's play a game. We'll have fun, and time will pass faster."

"I don't want to have fun," says the mathematician. "I want to get some sleep."

"But it's a nice game. I'm sure you'll like it." And the computer scientist explains, "I ask you a question. If you cannot answer it, you give me five bucks. Then you ask me a question, and if I don't know the answer, I'll give you five bucks."

The mathematician looks at the computer scientist as if the other man has two heads. "This is the stupidest game I've ever heard. It's not even properly formulated: the 'if' clause is only half defined—it doesn't say what happens in the case of a correct answer. Leave me alone. I want to sleep."

But the computer scientist has the bit between his teeth and doesn't let go. "OK, I'll make it really exciting for you. If you don't know the answer to my question, you give me five bucks; if I don't know the answer to your question, I give you a hundred bucks. What do you say?"

The mathematician realizes that the only way to shut up his fellow traveler is to give in and play the game, so he brings his seat to the upright position and, grudgingly, says, "Very well, ask your question."

The computer scientist beams. "What is the exact distance between the center of the Earth and the center of the Moon?"

"360,000 km," the mathematician answers.

"Nope," the computer scientist says. "That's an approximation. I want the exact figure."

"I don't know the exact figure," the mathematician says, and hands over a five-dollar bill.

"Your question now," the computer scientist says, very pleased with the turn of events.

The mathematician looks at him and asks, "What goes up the hill on five legs and down the hill on three legs?"

"Hmm...," the other muses. He whips out his laptop and searches the Encarta and other similar materials he has on it, but cannot find an answer. Then he connects through the plane circuitry to the internet and e-mails all his friends around the globe: no success. He tries other web sources, including the Library of Congress—still nothing. In the end, tired and annoyed, he nudges the mathematician (who, in the meantime, had got a good couple of hours of sleep). "I give up," he says, and hands over a hundred-dollar bill. The mathematician accepts the bill with indifference, puts it in his pocket, then turns his head away, intending to go back to sleep.

But the computer scientist, quite incensed, would have none of that. "Hey, wait a minute," he says irritably. "You must now tell me: what goes up the hill on five legs and down the hill on three legs?"

The mathematician turns his head back again and says, "Ah, so this is your second question." Then he produces another five-dollar bill and calmly gives it to the computer scientist. "I'm afraid I don't know the answer to that."

"Why a computer scientist?" JJ asked. "It could very well have been an engineer."

"Listen," I said. "Who's telling the stories: you or I? It was a computer scientist, okay? Now, here's the second one, which shows an innovative way of applying mathematical induction."[4]

---

[4] See p. 258.

A Boeing 747 is flying from London, England, to Houston, Texas. Two hours into the flight, the entertainment programs are interrupted by an announcement:

"This is your captain speaking. I have to tell you that engine number 1 has been malfunctioning and that, for safety reasons, I had to shut it off. However, there is no need to worry because the other three engines are fully capable of carrying us to our destination. The only inconvenience, for which we apologize, is that we'll get there two hours late."

The passengers look at each other with concern, then, trusting that the captain knows what he is doing, go back to their individual pursuits.

An hour later, a second announcement:

"This is your captain speaking. We're having an unlucky day today: engine number 2 has just stopped working and we can't restart it. So we'll have to continue our flight on the remaining two engines, which is perfectly safe, except that we will arrive at our destination four hours late."

The passengers again show concern, and again they have no alternative but to accept the captain's decision.

After another hour, a third announcement:

"This is your captain speaking. You won't believe it, but engine number 3 has also given up the ghost. However, since a 747 is fully capable of flying with a single engine, we have nothing to fear. The only problem is that we will be in Houston eight hours late."

Hearing this, a passenger, who has had one free alcoholic drink too many, says to a passing flight attendant, "Missh...I hope to GAWD, Missh, that engine number 4 is going to be OK, because if that one also conks out, we'll be flying up here for ever."

"Was this passenger an engineer?" JJ asked.

"I don't think even a drunk engineer would apply induction so eccentrically. Why? You want to talk about engineers? You've mentioned them twice already."

"Maybe some other time. You said you had three stories, didn't you?"

"Yes. And here's the third one."

Having survived the big flood, Noah's ark finds a shore and the old man releases all the animals on board into the wild. A few months later, he comes back to check on the animals' progress and is very pleased to see that they have all gone forth and multiplied, as he had instructed them. All, that is, except a pair of snakes, which were still just by themselves.

"What's wrong with you two?" he asks the snakes. "Why haven't you produced any young, like all the other critters? You live in a nice forest, have plenty of food, and no predators to worry about. What more do you want?"

"Cut down a few trees for us," the snakes hiss, "then cut them up. That should help us do your bidding."

Noah thinks that the snakes might have been affected by the tremendous amount of lightning and thunder during the big downpour, but decides to humor them and complies with their request.

A few more weeks down the line, when he returns to the forest, he sees that there are now lots of little snakes slithering around. He finds the original pair and asks, "What's the secret? What's so special about that cut-up timber? How in the world did that help you?"

"Simple," the snakes reply. "We're adders, so we need logs to multiply."

"Oh, yes, good old logarithms. They can fell an unwary student at ten paces." JJ put his hands on the table. "Thanks for the stories," he said. "I've really enjoyed our little chat. If all goes well, we'll talk some more next time." Then, unceremoniously, he sprang to his feet and promptly walked out of the bar.

Notes After the Meeting

Here are the answers to JJ's questions together with some brief personal comments.

Solution to Question 1 (arithmetic). The simplest way to find the answer is to remark that the new student's age is 1 year above the previous class average. Distributed equally over all the students now in the class, this difference creates an increase of 0.05 in the class average age. Therefore, the class currently consists of
$$\frac{1}{0.05} = 20 \text{ students.}$$

Solution to Question 1 (algebraic). Let n be the number of students after the new arrival. The total age of the class before was

(previous class average age) × (previous number of students) = $13(n-1)$.

The new student brings the total class age to $13(n-1) + 14$, which must equal

(current class average age) × (current number of students) = $13.05\,n$;

hence,
$$13(n-1) + 14 = 13.05\,n,$$
from which $n = 20$.

Comparing the two solutions, we should admit once again that the former is more elegant (and shorter) than the latter. Yet people don't normally think of solving the problem that way. Arithmetic is a much undervalued branch of mathematics, badly neglected in elementary schools, a victim of the I-have-a-calculator-so-why-should-I-use-my-brain doctrine. Shame, really....

SOLUTION TO QUESTION 2. Finite sums may be rearranged in this way; for example,
$$\begin{aligned} a - b + c - d + e - f &= (a-b) + (c-d) + (e-f), \\ a - b + c - d + e - f &= a - (b-c) - (d-e) - f; \end{aligned} \quad (16.3)$$

infinite series, in general, may not. We notice that, whereas in (16.3) there is a last term to take up the slack, so to speak, no such term exists in an infinite series. The series in Question 2 oscillates between -1 and 1 and is, therefore, divergent (it has no sum to infinity).

But even some series that do have a sum to infinity cannot be manipulated in this manner. Consider this one:

$$\sum_{n=1}^{\infty} \frac{(-1)^{n-1}}{n} = 1 - \tfrac{1}{2} + \tfrac{1}{3} - \tfrac{1}{4} + \tfrac{1}{5} - \tfrac{1}{6} + \tfrac{1}{7} - \tfrac{1}{8} + \cdots = \ln 2.$$

The more terms we add, the closer we get to the value of $\ln 2$, and if we add sufficiently many, we can get as close to $\ln 2$ as we wish; we say that the series sums up (or converges) to $\ln 2$. Let us assume that the terms can be rearranged, and do this in a special way:

$$\begin{aligned}
\ln 2 &= 1 - \tfrac{1}{2} + \tfrac{1}{3} - \tfrac{1}{4} + \tfrac{1}{5} - \tfrac{1}{6} + \tfrac{1}{7} - \tfrac{1}{8} + \cdots \\
&= \left(1 - \tfrac{1}{2}\right) - \tfrac{1}{4} + \left(\tfrac{1}{3} - \tfrac{1}{6}\right) - \tfrac{1}{8} + \left(\tfrac{1}{5} - \tfrac{1}{10}\right) - \tfrac{1}{12} + \left(\tfrac{1}{7} - \tfrac{1}{14}\right) - \tfrac{1}{16} + \cdots \\
&= \tfrac{1}{2} - \tfrac{1}{4} + \tfrac{1}{6} - \tfrac{1}{8} + \tfrac{1}{10} - \tfrac{1}{12} + \tfrac{1}{14} - \tfrac{1}{16} + \cdots \\
&= \tfrac{1}{2}\left(1 - \tfrac{1}{2} + \tfrac{1}{3} - \tfrac{1}{4} + \tfrac{1}{5} - \tfrac{1}{6} + \tfrac{1}{7} - \tfrac{1}{8} + \cdots\right) = \tfrac{1}{2}\ln 2.
\end{aligned}$$

This would imply that $1 = \tfrac{1}{2}$, another 'result' we reject indignantly, which means that the above rearrangement is illegal.

What type of series may be rearranged? One class for which this can be done consists of the so-called *absolutely convergent series*. A definition of these series can be found in any advanced calculus book.

As for the theorem we 'proved', it is clear that its statement can be trivially generalized to the ownership of a Ferrari or Maserati or any other prohibitively expensive car one may see oneself driving in one's dreams.

And we can extend the argument much further than that. By (16.2), whatever job we've got, it pays nothing, so we might as well stay unemployed. But we can have anything we want for free. If this 'alternative mathematics' worked, then we could live in a communist society, which would please Karl Marx[5] posthumously no end.

A Word to the Wise

Logarithms

 DO

The main properties of the (natural) logarithm are summarized below.

---

[5] German philosopher and political economist, 1818–1883.

$$\ln(ab) = \ln a + \ln b, \quad a,b > 0,$$
$$\ln \frac{a}{b} = \ln a - \ln b, \quad a,b > 0,$$
$$\ln(a^b) = b \ln a, \quad a > 0, \ b \text{ real},$$
$$\ln 1 = 0.$$

These properties remain valid if the logarithm is taken to any other base.

☠ **DON'T**

Here are some examples of objectionable manipulation involving logarithms:

- ☠ $\ln(ab) = (\ln a)(\ln b),$
- ☠ $\ln \dfrac{a}{b} = \dfrac{\ln a}{\ln b},$
- ☠ $\ln(a^b) = (\ln a)^b,$
- ☠ $\ln(a+b) = \ln a + \ln b.$

SCAM 17

The Public Media

> *I am a journalist and, under the modern journalist's code of Olympian objectivity (and total purity of motive), I am absolved of responsibility. We journalists don't have to step on roaches. All we have to do is turn on the kitchen light and watch the critters scurry.*
> Patrick Jake O'Rourke[1]

> *I hate journalists. There is nothing in them but tittering jeering emptiness ... The shallowest people on the ridge of the earth.*
> William Butler Yeats[2]

I met JJ again at a mathematics conference sponsored by Queens University in Charlotte, North Carolina. Just like the first time, he was sitting alone at a table in a corner of the hotel bar, gazing at me across the room with dark, inscrutable eyes.

As soon as I sat down, he pushed a piece of paper in my direction. "Here," he said laconically. "For your students."

I glanced at the paper and smiled. The text read something like this.

Question 1 (requires basic geometry). *What, if anything, is wrong with the following assertion and proof?*

Theorem. *All realtors are charlatans and swindlers.*

Proof. To avoid cumbersome notation, in what follows no symbolic distinction is made between an angle and its measure or between a line segment and its length, and congruence is denoted by the equality sign.

(i) First, we show that there is no angle with measure greater than $90°$. This is done by geometric construction. Draw a horizontal straight line segment AB and then another one AD perpendicular to AB (see Fig. 17.1).

Next, from B, draw a segment BC so that

$$BC = AD, \quad \angle ABC > 90°,$$

that is, BC makes an obtuse angle with AB. Join points D and C. Since $\angle ABC > \angle BAD$, DC is not parallel to AB; hence, the perpendicular bisec-

---

[1] American journalist and writer, b1947.
[2] Irish poet, dramatist, and public figure, 1865–1939.

tors of AB and DC, drawn through the midpoints P and Q of AB and DC, respectively, are not parallel and meet at some point O.

We claim that triangles OAD and OBC are congruent. This is indeed true because

(1) $OA = OB$ (OP is the perpendicular bisector of AB, therefore O is equidistant from A and B);
(2) $OD = OC$ (OQ is the perpendicular bisector of DC);
(3) $AD = BC$ (by construction).

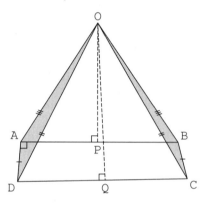

Fig. 17.1.

Since, in congruent triangles, corresponding angles are congruent, it follows that
$$\angle OBC = \angle OAD. \tag{17.1}$$
Also, triangle OAB is isosceles because $OA = OB$, so
$$\angle OBA = \angle OAB. \tag{17.2}$$
Subtracting equalities (17.1) and (17.2) side by side, we find that
$$\angle OBC - \angle OBA = \angle OAD - \angle OAB,$$
or
$$\angle ABC = \angle BAD = 90°;$$
therefore, any angle that appears to be greater than $90°$ is, in fact, $90°$.

(ii) Next, we show that any positive number is equal to 0. Let x be an arbitrary positive number. By part (i) of the proof, the angle $(90 + x)°$ is equal to $90°$, that is, $90 + x = 90$. Subtracting 90 from both sides, we find that
$$x = 0. \tag{17.3}$$

(iii) Consider a realtor and suppose that he claims to be able to sell a house for $x. By (17.3), he is, in fact, incapable of getting more than $0 for it, so he is a charlatan. Invoking his contract with the seller, he then demands a fee of 7% of x. Again by (17.3), all he is entitled to is 7% of 0, in other words, absolutely nothing. But he won't accept that and grabs $0.07\,x$, which makes him a swindler. Since this realtor was picked at random, we conclude that all realtors are charlatans and swindlers.

Question 2 (requires basic geometry). *Suppose that Earth is a perfectly smooth sphere. An inextensible string is tied around the equator (a 'great circle' on the sphere, $40,000,000$ m long) in such a way that it makes contact with the surface at each point. Another piece of 1 m is then added to the original length, and the new, longer, string is formed into a circle with a uniform clearance above the equator. Can a normal-size cat crawl under the new string without touching it?*

"Interesting," I said. "Thanks." I folded the paper and put it in my pocket, waiting for JJ to broach, as usual, a subject of his choice. He needed no special invitation.

"You've got a big problem with the media on this planet," he said abruptly.

Was it possible that we could, at last, completely agree on something? Still unsure, I decided to play it safe. "What have you got against them?"

"There are individuals in the media who, like some defense lawyers, warp logic to make it justify their perversion of truth and unethical behavior. Not so long ago, Gan completed a VOTSIT survey of reporters. I'll let you digest the answers and judge for yourself. To keep things simple, as usual, we'll give the respondents the same generic name, Rep."

GAN: Why do you almost always report what's bad and hardly ever what's good?"

REP: What's good is bland and unexciting. What's bad shocks the viewers, listeners, and readers, who will want more of it. Good is boring. Bad is fascinating because it feeds the consumers' limitless appetite for the morbid.

GAN: Do you pick your topics according to the public's preferences?

REP: I don't care what people like. *I* tell them what they like. I'm an opinion former.

GAN: Is your reporting always truthful?

REP (dismissively): Truth is much overrated. Besides, it's relative. I keep to the truth just enough to avoid libel charges. Most often, news is flat and soporific, so I spike it up or embellish it a little. Sometimes, even a little more.

GAN: To make it interesting?

REP: To make it sensational. Ours is a cut-throat business. We do whatever is necessary to stay ahead of the competition.

GAN: A while ago, you claimed that a young man had murdered his parents, and backed your claim with totally inconclusive 'evidence', part of it untrue. The young man was then relentlessly harassed by you and your colleagues until, unable to take it any more, he committed suicide. Did you not feel guilty when the police caught the real criminal a few weeks later?

REP: 'Guilty'? What does that mean?

GAN: During one of your investigations, you detected a weak point in the national security measures. But instead of quietly bringing the facts to the attention of the authorities, you broadcast them all over the airwaves. Didn't you think that this might help the enemy?

REP: The enemy is our government's concern, not mine. I have a duty to let the public know what's going on, and show them how clever I am.

GAN: If, when next time you do this again, the enemy makes quick use of the information supplied by you and takes action that results in the loss of thousands of innocent lives, would you still maintain that what you did was right?

REP: Absolutely. The people who die will do so fully informed.

GAN: When someone suffers a terrible personal tragedy, why do you, journalists, descend on him like a pack of hyenas, shove mikes in his face, and ask, "How do you feel?" What do you expect him to say? That life's great and he's enjoying every minute of it?

REP: I don't give a toss what he says or feels. All I want is to be caught on camera and seen fighting on the front line. This helps me make a name for myself and move forward in my career.

GAN: You write articles about 150-year-old women giving birth to three-legged monsters, dinosaurs kept in someone's backyard as pets, discovery of a tunnel joining the North Pole with the South Pole through the center of the Earth, and other things like that. Is there a readership for such drivel?

REP: I wouldn't do it if there weren't. Some people will believe anything. My articles won't get me the Pulitzer Prize,[3] but at least I earn a comfortable living."

GAN: Are you not ashamed that, instead of helping educate the masses, you are feeding them something that numbs their minds and stunts their cultural growth?

REP: I'm their intellectual fast food provider. I give them something they can swallow quickly, something that makes them think they are getting the right stuff, but which, as I am well aware, is anything but. Do I care that they get short-changed? Do I *look* like I care?

"Some time ago," JJ said, "by way of experiment, Gan arranged to have a spoof item inserted in one of the tabloids. Take a look."

Shock Discovery: Numbers Are Alive!

After many years of deep and intense study, a team of high-power mathematicians led by Dr. Philip Schabernack[4] at the Center for the Humane Treatment of Science has proved that numbers are, in fact, live creatures populating a universe parallel to our own. They are organized in groups where some rise to a position of power, but their associative structure has a ring of mystery about it for strangers to the field; they talk to each other and to us through the language of arithmetic; they always tell the truth, although we fail on occasion to grasp it; and they also

---

[3] A much coveted American award in newspaper journalism, literature, and music, named after Joseph Pulitzer (Hungarian-born American journalist and publisher, 1847–1911).

[4] Practical joke; hoax. (German)

have gender and experience pairwise multiplication. Referring to positive integers, Dr. Schabernack said, "It is obvious that odd numbers are male and even ones are female. The lack of parity of the former mirrors the single-mindedness and egotistic nature of men, who like to think they are number 1, whereas the factor of 2 present in the latter implies clear accents of femininity. Besides, when that factor is pulled out, the set of even numbers generates all positive integers, a reproductive property the odd contingent does not have." Dr. Schabernack went on to say that, since numbers enjoy full immortality, the government should pour money into further research to learn how we, humans, could also live forever.

"Did anyone take this seriously?" I wondered.

"After publication, the newspaper got a couple of calls from readers who wanted Congress to give guarantees that, if mathematicians discovered the secret of immortality, they should not be the only ones to profit from it. What comes out of taxpayers' money must be shared by all."

"I've seen plenty of bad journalism myself," I said, "but you can't deny that there are many decent, highly professional reporters around."

"Not nearly enough. Too many unBRITEs still among them, whose lack of social and moral responsibility in the name of the First Amendment compromises the guild's integrity. I see this as I see a diseased body: you have a bunch of healthy cells being stifled by a larger bunch of bad ones. And then," JJ added, "you've got the movies."

"So Gan interviewed movie people, too..."

"He did. For your benefit, here is a sample of what a typical movie producer, Mov, had to say."

GAN: In one of your films, you completely distorted the historical facts and attributed a foreign invention to one of your conationals. Why did you do that?

MOV: You think our viewers would spend their cash to see us singing the praises of a foreigner? This is Hollywood! We do as we please, so long as we rake in the money.

GAN: You made a movie in which alien invaders of Earth, possessed of a technology that allowed them to travel faster than light across the galaxy, are defeated by a computer virus delivered with a laptop. Aren't you insulting the intelligence of the viewers?

MOV: Such films are not made for people like you. They are for those who go to the movie theater mostly to munch popcorn, drink soda, be deafened by the speakers, stamp their feet, and talk loudly to each other. They don't care about plausibility. They would have no problem even if the virus were delivered by blowpipe.

GAN: A scene in one of your movies had some mathematical formulas displayed on a classroom board. To the expert eye, those formulas were complete gibberish. Why didn't you call on a college professor to write for you a bit of correct calculus?

MOV: Calculus, shmalculus. Look at the statistics. The movie in question has been watched by 20 million people. Of these, 5,000 are math teachers, and 4,500 noticed that the formulas were meaningless, but were not bothered by the fact. The great majority of the remaining 19,995,000 have gone through our public school system and face massive problems with long division. They don't even know what calculus is.

GAN: In one of your movies, a character was heard using a Latin phrase. The correct expression is *Si vis pacem para bellum*.[5] However, when it was flashed across the screen—to help the viewers, I suppose, and also add a flavor of culture to the production—its first word was given as *sic*, which means 'thus', not 'if'. Why this cavalier, semidoct attitude?

MOV: Well done. You are one of five viewers who noticed the error.

GAN: Do you know what 'infinity' means?

MOV: Of course. It's what we are aiming for in our box office takes.

GAN: Then what is $\infty - \infty$?

MOV (without hesitation): Zero.

GAN: I thought you might say that. You are treating infinity as a number. This is the answer of someone with a MICQ no higher than 19, who never heard of the Fourth Mathematical Commandment.

"People like these Rep and Mov," JJ said, "who mislead and deceive, are keeping your planetary MICQ well below the critical 50 level."

"Any views on television?" I asked.

"Let's not talk about that. Many of your TV networks specialize in trash, fit for an audience marginally smarter than a kindergarten dropout. There is something fundamentally wrong with the cultural health of a country where decent programs have to be produced or bought from public donations." JJ leaned back in his chair. "Now," he said, "it's time for your stories."

I finished the last of the coffee and pushed the empty cup aside. "OK, here we go. Today I've got a couple touching on the media and a third involving numbers. The first one is not unlike Gan's earlier spoof, except that it's based on a real incident."

Front page article in the local newspaper, Monday, August 25:

Math Professor Brings Tennis Ball to Class

On the first day of the new academic year, Dr. Pom, a professor of mathematics at the City State University, placed a tennis ball on his desk in full view of the students, balancing it carefully to prevent rolling. The students expected the ball to be used as a prop for some problem during the lecture, but were completely bemused when this did not happen. The consensus was that, like many math professors, their instructor was somewhat odd.

Front page article in the local newspaper, Friday, August 29:

Tennis Ball Professor Keeps Students Guessing

Dr. Pom continues to bring the now famous tennis ball to every class, still without an explanation. His intrigued students have started speculating variously that the

---

[5] If you want peace, prepare for war.

ball might be (i) an indication of the prof's passion for tennis, (ii) a reminder of some great, inspirational win on court, (iii) a symbol of the mathematically perfect shape, the sphere, (iv) a statement that mathematics, being part of life, is ultimately just a game, and so on.

Front page article in the local newspaper, Monday, September 1:

Classroom Mystery Solved as Prof Makes Point but May Lose Game

The meaning of the tennis ball on Dr. Pom's desk was revealed today by the doc himself in spectacular fashion. During his morning lecture, one of the students fell asleep and began snoring loudly. Calm and composed, the prof stopped his explanations at the board, took aim, and bounced the ball with extraordinary precision off the offender's head. He was subsequently reported to the dean and accused of cruel and unusual punishment. Although this is reason for dismissal, the great majority of Dr. Pom's students hope that the dean will act leniently. They signed a petition which claims that the snoring in question had been particularly unpleasant in both pitch and volume, that the doc's fully justifiable reaction met with their approval, and that it would be good if all their instructors brought tennis balls to their lectures.

Front page article in the local newspaper, Wednesday, September 3:

Doc Pom Back in Class, Earns New Moniker

The prof who threw a tennis ball at a snoring student on Monday reappeared before his classroom audience this morning, having gotten away with just an admonishment from the dean. As soon as the doc displayed his paraphernalia on the desk, the students came up with a new nickname for him: the Baseball Professor...

"After which, in the fullness of time, he might become the Shot-put Doc?" JJ said.

"Second story coming up."

A tabloid reporter on a country road glimpses a little boy in a field being charged by an enraged bull. The reporter stops the car, whips out his camera, and starts shooting pictures. As the boy, rooted to the spot, is watching the fast-approaching beast with eyes full of terror, a second car pulls up. Its driver climbs over the barbed wire fence, rushes to the scene, grabs the bull by the horns, wrestles it to the ground, and, with a sharp twist of its head, breaks its neck. He then turns to the frightened child and comforts him.

When the man returns to his car, the reporter, who has photographed every moment of the action, shakes his hand enthusiastically and says, "Sir, I'm privileged and proud to meet a true American hero!" To which the stranger, sketching an embarrassed smile, replies, *"Mi dispiace, ma non parlo inglese."* [6]

On the front page of next morning's edition of the tabloid, above the picture of the tourist hugging the little boy by the dead animal, the headline proclaims in big, fat letters: "Italian Pedophile Kills Child's Pet."

---

[6] I'm sorry, but I don't speak English. (Italian)

JJ seemed to chuckle. "How unusual! A newspaperman who not only falsifies the truth, but for whom recording the goring of a child is more important than trying to rescue him. Gan will love this one."

"Now," I said, "the third and final offering."

A group of reporters and photographers are passing the time before the start of a press conference.

"249!" one of them shouts, and everybody else bursts into loud guffaws.

As the noise dies down, another one raises his hand and says, "413!" The response is big grins and some polite laughter.

At this point, a novice joins the group and quietly asks a veteran, "What's going on? What are these numbers and why does everybody laugh at them?"

The man explains that since time is at a premium for workers in their profession, a while back they had compiled a list of all the jokes they knew, assigned numerical codes to them, and, ever since, instead of wasting time telling a joke, they just call its number.

"Great idea," the young man agrees. Then, on impulse, he decides to have a go and shouts, "364!"

When nobody reacts, he tries another: "157!" Again, not a sound. The newcomer resolves to make one last attempt and remembers the number that had made everybody laugh their heads off. "249!" he says with gusto, throwing in a wink for good measure. The others are just staring at him in complete silence.

"What's wrong? Why doesn't anyone laugh?" he asks the veteran.

The older man shakes his head and says, "It's the way you tell them. You just don't have the flair for it, sonny."

"Some journalists should use the same encoding procedure with their pieces and just call them out by their numerical label. This would be less boring and offensive for their listeners or readers." JJ put his hands on the table. "Thanks for the stories," he said. "I've really enjoyed our little chat. If all goes well, we'll talk some more next time." Then, unceremoniously, he sprang to his feet and promptly walked out of the bar.

Notes After the Meeting

Here are the answers to JJ's questions together with some brief personal comments.

Solution to Question 1. As mentioned earlier, any mathematical proof consists of three main parts: a premise (here, a geometric construction), a logical argument, and a conclusion. The conclusion is based on the argument, which, in turn, is based on the premise. So, when the argument is correct but the conclusion is wrong, we must assume that the premise is specious.

Generating the construction by computer, we immediately see that Fig. 17.1 is incorrect and that the rigorous version is as shown in Fig. 17.2.

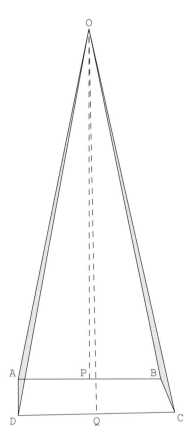

Fig. 17.2.

The crucial difference between Fig. 17.1 and Fig. 17.2 is the position of triangle OBC. The mathematical argument does not change: we still find triangles OAD and OBC to be congruent, and we still conclude that ∠OBC and ∠OAD have the same measure, but this tells us absolutely nothing about the measure of ∠ABC.

A while ago, I showed the 'proof' in part (i) to some of my engineering colleagues during lunch at the faculty club. Naturally, they didn't buy the phony conclusion that all obtuse angles are right angles, but then barked up the wrong tree, attempting to find a flaw in the mathematical argument. After a few minutes of sterile discussion, one of them, a young research assistant, quietly left and came back a little later with a computer printout showing the correct figure, which he slammed triumphantly in the middle of the table.

"There!" he said with great relief. "That's where you went wrong!"

"I didn't go wrong anywhere," I answered. "I knew exactly what I was doing. One important fact you should learn about mathematicians is that, whether the lines they draw are straight or wobbly, their reasoning is always straight."

As mentioned in the first paragraph of the 'proof' in Question 1, sometimes mathematical language can become cumbersome. For example, two distinct geometric objects (line segments, angles, triangles, etc.) that have the same shape and size are called *congruent*, not equal. Thankfully, ordinary English does not suffer from such complications. If it did, then Thomas Jefferson[7] might have written in the Declaration of Independence that all men are created 'congruent' instead of 'equal'.

Solution to Question 2. Let R be the radius of Earth's equator (hence, also of the original string) and R' the radius of the new circular string positioned above the equator. Then the lengths L and L' of the two concentric circles in Fig. 17.3 are, respectively,

$$L = 2\pi R = 40{,}000{,}000 \text{ m},$$
$$L' = 2\pi R' = 40{,}000{,}001 \text{ m}.$$

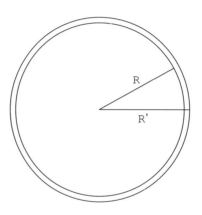

Fig. 17.3.

Subtracting these equalities side by side, we find that

$$L' - L = 2\pi(R' - R) = 1 \text{ m}, \qquad (17.4)$$

from which we compute that the uniform clearance of the new string above the surface is

$$R' - R = \frac{1}{2\pi} \cong 0.16 \text{ m},$$

or 16 cm. A normal-size cat is perfectly able to crawl through a gap this narrow.

When an audience is asked to guess the answer, a few people say 'no', a few say 'yes', and the majority say nothing at all.

---

[7] The third American president, 1743–1826.

The first type of responder acts on intuition, which, in view of the huge difference between the length of the original string and the tiny addition to it, reckons that the clearance must be too small for a cat to crawl through. Although wrong, these responders should be commended on their honesty.

The second type presumably comes to the same intuitive conclusion, but then goes a step further and argues that if the answer were so *obvious,* the question would not have been asked at all, so they conclude that there must be a catch somewhere and give the opposite answer.

The third type of responder is the least appreciated. Sitting on the fence is not conducive to solving mathematical problems. In mathematics we must take chances, must 'declare' a direction of search, must try to develop a feeling for the result, and must work tenaciously until either the task is completed or the chosen road dead-ends, in which case, after licking our wounds and identifying where the quest failed, we adjust our course and start battling anew.

In the problem at hand, because the clearance $(R' - R)$ depends only on the added length $(L' - L)$, and because this dependence is linear, the clearance remains the same if we assume that the entire universe is a sphere and we tie a string around it, which we subsequently lengthen by 1 m. However large the sphere is, the answer doesn't change. This is one case where intuition is proved wrong by mathematics.

A direct application of this result, outside the exciting but rather limited-interest world of cats and strings, is in assessing the length of the flight path of an airliner. How much longer will the path be if, to avoid turbulence, a passenger plane flying around the world needs to increase its altitude by 300 m?[8] The first equality in (17.4) with $R' - R = 300$ m yields

$$L' - L = 2\pi(R' - R) \cong 2 \times 3.14 \times 300 = 1,884 \text{ m} \cong 1.9 \text{ km},$$

which is very little. The main concern of the pilot in this situation is not, therefore, the extra distance, but the amount of fuel necessary to climb to the higher altitude. If the plane is originally flying at, say, 11,000 m, then a rough percentage estimate of the extra fuel required by the maneuver is

$$\frac{300}{11,000} \times 100 \cong 2.7\%.$$

This might seem like a lot for a flight around the globe, but less so for just an intercontinental hop. A supplementary—but much smaller—amount of fuel is also necessary to make the additional descent at landing.

A Word to the Wise

Geometry

A triangle with two congruent sides (angles) is called isosceles; one with all three sides (angles) congruent is called equilateral.

---

[8] Approximately 1,000 ft.

The sum of the measures of the interior angles of a triangle is 180°, or π radians.

Here are the formulas for computing some measurable characteristics of a few geometric objects.

Area of a triangle:
$$A = \tfrac{1}{2} bh,$$
where b is the length of the base (any one of the triangle's sides) and h is the length of the corresponding height (altitude).

Area of a rectangle:
$$A = ab,$$
where a and b are the lengths of two adjacent sides.

Area of a trapezoid:
$$A = \tfrac{1}{2}(a_1 + a_2)h,$$
where a_1 and a_2 are the lengths of the two parallel sides and h is the distance between them.

Length and area of a circle of radius r:
$$L = 2\pi r, \quad A = \pi r^2.$$

Volume of a rectangular box:
$$V = abc,$$
where a, b, and c are the lengths of the three edges joining at any vertex.

Volume of a pyramid:
$$V = \tfrac{1}{3} Ah,$$
where A is the area of the base and h is the height (the volume of a cone is given by the same formula).

Surface area and volume of a sphere of radius r:
$$S = 4\pi r^2, \quad V = \tfrac{4}{3}\pi r^3.$$

In analytic geometry the general equation of a straight line in the Cartesian (x, y)-plane is
$$ax + by + c = 0, \quad a, b \text{ not both zero.}$$

The general equation of a circle is
$$x^2 + y^2 + ax + by + c = 0, \quad \text{where } a^2 + b^2 \geq 4c.$$

In terms of the coordinates (x_0, y_0) of the center and its radius r, the equation of the circle takes the form
$$(x - x_0)^2 + (y - y_0)^2 = r^2.$$

SCAM 18

The Criminal Legal System

> *Aucun poète n'a jamais interprété la nature aussi librement qu'un avocat interprète la réalité.*[1]
> Jean Giraudoux[2]

> *Who thinks the law has anything to do with justice? It's what we have because we can't have justice.*
> William McIlvanney[3]

I met JJ again at a mathematics conference sponsored by Rice University in Houston, Texas. Just like the first time, he was sitting alone at a table in a corner of the hotel bar, gazing at me across the room with dark, inscrutable eyes.

As soon as I sat down, he pushed a piece of paper in my direction. "Here," he said laconically. "For your students."

I glanced at the paper and smiled. The text read something like this.

Question 1 (requires elementary logic). *The population of Funnyland*[4] *consists only of Truth-Tellers (TTs) and Pathological Liars (PLs). A foreigner who has applied for the position of court clerk in the capital city is told that he must pass a preliminary test before being considered for the job.*

At the appointed time, the applicant, V, is shown into a room where three Funnylanders, D, H, and Q, are sitting behind a table, and is handed a piece of paper that details his assignment, as follows: "The panel in front of you consists of two defense attorneys and a judge. You are allowed to ask a single question, which will be answered by each panel member. Analyze the answers and identify the judge and the defense attorneys."

V ponders for a few seconds, then asks, "Which one of you is the judge?"

Calmly, D replies, "I am."

Equally calmly, H says, "I am."

Completely unperturbed, Q advises, "Sir, think very carefully because at most one of us is a TT."

V thinks very carefully and determines who is who. How did he do it?

---

[1] No poet ever interpreted nature as freely as a lawyer interprets truth. (French)
[2] French dramatist, 1882–1944.
[3] Scottish novelist, b1936.
[4] See Question 2 on p. 34.

Question 2 (requires basic trigonometry). What, if anything, is wrong with the following assertion and proof?

Theorem. *The law is an ass.*

Proof. Once again, to avoid complicating the notation, in what follows no symbolic distinction is made between an angle and its measure or between a line segment and its length, and congruence is denoted by the equality sign.

(i) First, we demonstrate that every triangle is isosceles, that is, it has two congruent angles (or, equivalently, two congruent sides).

Consider a triangle ABC, which, to make matters credible, would appear not to be isosceles. Extend AB beyond A by a segment $AS = AC$ and AC beyond A by a segment $AR = AB$ (see Fig. 18.1).

By construction, triangles ABR and ACS are isosceles, with

$$\angle ARB = \angle ABR, \quad \angle ASC = \angle ACS.$$

Since

$$\angle ARB + \angle ABR = \angle BAC, \quad \angle ASC + \angle ACS = \angle BAC,$$

we see that

$$\angle ARB = \angle ABR = \tfrac{1}{2}\angle BAC, \quad \angle ASC = \angle ACS = \tfrac{1}{2}\angle BAC. \tag{18.1}$$

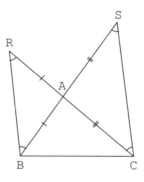

Fig. 18.1.

By the sine theorem in triangles BCR and BCS, respectively, and (18.1),

$$\frac{CR}{\sin \angle CBR} = \frac{BC}{\sin \angle CRB} = \frac{BC}{\sin\left(\tfrac{1}{2}\angle BAC\right)},$$

$$\frac{BS}{\sin \angle BCS} = \frac{BC}{\sin \angle BSC} = \frac{BC}{\sin\left(\tfrac{1}{2}\angle BAC\right)},$$

from which it follows that
$$\frac{CR}{\sin \angle CBR} = \frac{BS}{\sin \angle BCS}. \qquad (18.2)$$

But
$$CR = AC + AR = AC + AB, \quad BS = AB + AS = AB + AC,$$

so $CR = BS$. Then (18.2) leads to
$$\sin \angle CBR = \sin \angle BCS. \qquad (18.3)$$

Now, by (18.1),
$$\angle CBR = \angle ABC + \angle ABR = \angle ABC + \tfrac{1}{2}\angle BAC,$$
$$\angle BCS = \angle ACB + \angle ACS = \angle ACB + \tfrac{1}{2}\angle BAC.$$

Consequently, (18.3) becomes
$$\sin\left(\angle ABC + \tfrac{1}{2}\angle BAC\right) = \sin\left(\angle ACB + \tfrac{1}{2}\angle BAC\right), \qquad (18.4)$$

implying that
$$\angle ABC + \tfrac{1}{2}\angle BAC = \angle ACB + \tfrac{1}{2}\angle BAC,$$

or
$$\angle ABC = \angle ACB.$$

This means that triangle ABC is isosceles.

(ii) Next, we show that 1 is the only positive integer. Without loss of generality, consider a triangle ABC with angles of $40°$, $60°$, and $80°$. (Recall that the sum of the measures of the interior angles of a triangle is always $180°$.) By part (i), two of the angles must be equal, so, again without loss of generality, we may assume, for example, that
$$60 = 40.$$
Subtracting 40 from both sides, we arrive at the equality
$$20 = 0,$$
which, on division by 20, yields
$$1 = 0. \qquad (18.5)$$
If we multiply this by any positive integer n, we find that
$$n = 0, \qquad (18.6)$$
which, combined with (18.5), leads to
$$n = 1. \qquad (18.7)$$

(iii) We now go over to the assertion of the theorem. Among the myriad things the law concerns itself with is polygamy. Suppose that a man has n wives. According to (18.7), he has, in fact, only one wife, which means that polygamy is an illusion. Therefore, the law makes a big fuss about, and stipulates punishment for, an imaginary crime, so we are forced to conclude that the law is an ass.

"Interesting," I said. "Thanks." I folded the paper and put it in my pocket, waiting for JJ to broach, as usual, a subject of his choice. He needed no special invitation.

"You want to improve the planetary MICQ," he started abruptly, "but you won't have any noticeable success as long as your criminal legal system is allowed to remain in its present shape."

"You mean, our criminal *justice* system?"

"I can't call it that. I would, if it operated on sound logic and delivered justice. But all too often this is not the case."

"What do you think is wrong with it?"

"To begin with, it's adversarial. You have two lawyers who slug it out in court, trying to outsmart each other. You have a judge who referees the fight, ensuring that the rules of procedure are adhered to but contributing hardly anything else to the show. And you have twelve jurors who, whether they understand what's being said or not, end up many times making decisions that baffle everybody, including themselves."

"You may not like the system, but it works."

"*Any* system can be made to work. The question is, does it work *well*? In my view, this one doesn't. Built on the total repudiation of the Sixth Mathematical Commandment, it's a very unBRITE institution."

"That's a pretty severe indictment. Do you have any proof?"

"Plenty. We've surveyed criminal lawyers and been startled to see how revealing their answers are when the VOTSIT is fully on. Here is what my colleague Gan got out of some defense attorneys, generically named Att."

GAN: What is your greatest concern in court? Establishing the truth?

ATT: Certainly. The truth that I concoct and sell to the jury. The truth that saves my client's bacon.

GAN: Not the absolute truth?

ATT (laughs heartily): The *absolute* truth? Be serious! Many times I *know* my client is guilty. But that makes no difference to me. All I want is to show everybody how clever I am, and that I can get the scoundrel off the hook in spite of the overwhelming evidence against him, even if it means harming some innocent people in the process.

GAN: This sounds cynical.

ATT: To you maybe. Not to me. My goal is well defined: I want to win the case no matter what. The more of a slimeball the client is, the greater my triumph.

GAN: Don't you feel remorse for putting murderers back on the street?

ATT: Remorse?! Why? First, they won't kill me because they need me to defend them. Second, if they kill again, then I'll get more business. No, I have no compunction doing what I'm doing. It's all about becoming famous and getting paid lots of money.

GAN: Aren't you afraid that the jury might see through your tricks?

ATT: When the jury is empaneled I do my best to make sure that anybody with a decent brain is thrown out. Those who finally get selected are putty in my hands. I can confuse them so thoroughly they won't even remember their names. I can tell them anything I want, and they'll believe me. I'm a very convincing actor.

GAN: Some would say that you are no different from your client because both of you are harming society.

ATT: Not true. There is a fundamental difference between us: he does it outside the law, I do it within the law.

GAN: If you were a juror who really wanted to see justice done, what would you do?

ATT: I wouldn't listen to a single word the defense attorney is saying. I would pay attention only to the prosecution and then make up my mind whether they have proved their case or not.

"In our eyes, as neutral observers," JJ commented, "your legal setup is deeply flawed. A properly structured judiciary shouldn't give murderers the chance to kill again, or make it possible for rapists to ruin more lives, or free pedophiles so they could go on molesting children."

"Before we convict anybody, we need to be sure that we've got the right person—"

"Oh, really? What about having the criminal clearly identified and letting him go scot-free on a technicality? A hundred witnesses see the guy plunge a knife in the victim's back, yet he walks because in the heat and chaos of the scene a rookie policeman did not say the prescribed words to him, or in the correct order, or delayed saying them by a couple of minutes."

"The Constitution guarantees the rights and freedoms of every individual."

"Very commendable. But an individual also has obligations. For example, the obligation not to cause unprovoked physical harm to another individual. The Constitution keeps remarkably quiet about such things. It also fails to proclaim the right of the victim not to be hurt at the hands of others. You've biased the system so much in favor of the criminal that it is threatening to turn into an irrelevancy. As Adam Smith[5] put it, "Mercy to the guilty is cruelty to the innocent." Yes, your Constitution is a great document. But it's an *old* document. The founding fathers wrote it for a society fundamentally different from yours. If you don't allow it to keep pace with the times, it will slowly become more of a hindrance than help."

"Has Gan ever discussed these issues in the classroom?"

---

[5] Scottish moral philosopher and one of the pioneers of political economy, 1723–1790.

"You bet. One particular conversation was very animated, when Stu asked him to find an analogy between the operational processes in the legal system and mathematics. I have the recording for you."

GAN: Logical reasoning shows that the practice of law may, in a certain sense, be accused of misrepresenting its aims. Consider the following chain of statements:
Justice cannot be administered unless the truth is known.
Defense attorneys do not normally care about the truth.
Defense attorneys form a large part of the legal system.
Therefore, to a large extent, the legal system is not concerned with justice.

STU: Don't get me wrong, Dr. Gan, I'm not a fan of lawyers, but I can't help thinking that their sources and tools are a lot less clear-cut than ours. When they embark on a case, they don't always have the luxury of knowing that one plus one is two.

GAN: This is true. However, even when they do know it, many of them abandon honesty and integrity and argue that one plus one may not be two because their clients paid them to do so.

STU: Can you give us an example?

GAN (works the projector-linked computer): Suppose that the prosecution's case rests on the statement

$$1 + 1 = 2. \qquad (A)$$

Addressing the jury, the defense attorney, Att, says that (A) can be expressed in the equivalent form

$$\ln e + \cosh^2 y - \sinh^2 y = \sum_{n=0}^{\infty} \frac{(\csc x)\sqrt{1 - \cos^2 x}}{2^n}. \qquad (B)$$

It is almost certain that the jurors don't have enough background in science, so they make no sense of (B) and are primed to accept anything this learned person is telling them. Att takes full advantage of the situation and applies the *coup de grâce*:[6] he says that, while (A) can be an explanation for (B), it is not the only possible one. He points out that $\sqrt{1 - \cos^2 x}$ has, in fact, two different values, namely $\sin x$ and $-\sin x$, and that the latter, in no way less valid than the former, leads to

$$1 + 1 = -2.$$

This, he concludes, casts reasonable doubt on the prosecution's case, which means that his client must be acquitted.

STU: An ingenious piece of sophistry, but totally laughable. Att applies double standards: he uses $\sin x = \sqrt{1 - \cos^2 x}$ when he builds up his 'argument', then conveniently introduces the $\pm$ in front of the root when he deconstructs it. He mentions nothing about the essential role played by the value of x in that formula and wrongly claims that (A) and (B) are equivalent.

---

[6] The final, victorious blow; literally, *blow of mercy*. (French)

GAN: You see now why Att would not want to have someone like you on the jury?

STU: Earlier, you said that the Constitution needs to keep pace with the social realities of today's world. Has mathematics ever experienced similar changes?

GAN: The science of imagination and precision is in a continuous state of flux. And when it readjusts, it does so not because it has been wrong, but because it is eager to adopt a wider and richer perspective, to embrace new views and concepts as they come on line. Yes, it may be said that mathematics has its own kind of 'constitution', whose 'articles' are a number of axioms, simple truths that human intuition accepts as obvious—well, most of the time—and from which all other results are derived. Consider geometry. For centuries, its study had been based on five of these postulates, formulated by Euclid.[7]

STU: I read somewhere that the fifth made mathematicians uncomfortable...

GAN: Indeed. Known as the parallel postulate, it states roughly that through a point not on a given line, one and only one line can be drawn that never meets the given line. In the 1800s, however, non-Euclidean geometries were developed, where the fifth postulate is not satisfied. Mathematics did not reject these geometries because they violated the existing 'constitution', but admitted them into its fold at once, which was just as well, since, on a cosmic scale, our universe turns out to be non-Euclidean.

STU: I think I understand what you are saying, Dr. Gan. Things that have served us well for a long time need to evolve if they are to remain useful. Unfortunately, lawyers are not mathematicians. They will oppose any change they see as detrimental to the size of their earnings.

"An insightful remark," I said. "Still, talking about criminals and rights, at least here you are allowed to defend yourself when attacked."

"That may be true in certain parts of the country. In others, you are permitted to respond only in a manner 'commensurate' with the action of the assailant. Suppose you are woken up in the middle of the night by a noise in the house. You grab a baseball bat and go downstairs, where you come face to face with a masked intruder holding a gun in his hand. How are you expected to react? Imagine the following scene between a house owner, Own, and a burglar, Bur.

OWN: Do you intend to shoot me, or are you using the gun only as a threat?
BUR: Huh? Give me your money!
OWN: Not before we establish clear rules of engagement. I need to know how to respond. What if I suggest that we may resort to a degree of violence not exceeding hitting each other over the head? Would that be acceptable to you?
BUR (motioning impatiently with his gun): The money, I said!
OWN (puts the bat down and whips out a piece of paper and a pen): Let's draw up a contract. 'I, the Burglar, hereby declare that I will not shoot the House Owner in my attempt to break in and steal his valuables, and that the most physical

---

[7] Greek mathematician, c325–c265 BC.

harm I'm prepared to do to him in the course of said action is hitting him over the head with my gun. I, the House Owner, hereby declare that I will not use any physical force against the Burglar while he is on these premises attempting to rob me, beyond hitting him over the head with my bat.' I'm going to sign and date it.... There! Now it's your turn.

BUR (lowers the gun, advances toward Own, hits him hard over the head, steps over his unconscious body, and starts ransacking the house): Sorry, buddy, but there is no clause in the contract as to *when* I'm allowed to hit you over the head. Man, don't you just love it when homeowners play it by the rules! Thank you, legislators, for making my life so easy.

"How silly is that?" JJ said. "If you found this armed hoodlum violating the sanctity of your home and threatening your life, wouldn't you see red and crack his head open before he had time to do anything?"

I didn't know what to answer, so I mumbled, "My response will be properly considered by a jury."

JJ seemed amused. "A jury. Yes. Recently, Gan completed a VOTSIT survey of people who served on juries. You might find it interesting to hear what a typical juror, Jur, had to say."

GAN: Did you understand the details of the case?

JUR: Some of them. I didn't concentrate all the time. Too much to take in, too quickly.

GAN: Then how could you decide that the defendant was not guilty?

JUR: Not enough evidence.

GAN: He shot a man on the field of play, in front of sixty thousand spectators. What more evidence do you need?

JUR: I wasn't one of them. I didn't see him pull the trigger. I had a reasonable doubt. Not guilty.

GAN: The prosecution brought tons of corroborating witnesses. The defense was pathetic. Why did you *really* side with the accused?

JUR: Look—my life is plain and boring, so when this chance to show that I matter came along, I took it and gave the state a bloody nose. Sure, the guy in the dock was guilty, but I didn't care. Besides, the defense attorney was very handsome.

GAN (scribbles on a piece of paper $(a+b)/a$): Can you simplify this?

JUR (crosses off a in the numerator and a in the denominator and writes 'b'): Here it is. Why?

GAN: I merely wanted to confirm that your MICQ is no higher than 9.

I pursed my lips. "All right, so you dislike our system. Tell me then: how do you deal with crime on *your* world?"

"Our crime rate is negligible. On the rare occasion that an illegality is committed, we find the truth and act accordingly."

"You *find* the truth?"

"I know this is an alien concept to you, but that's exactly what we do. We don't operate with ifs and maybes; we use the facts themselves. The person under suspicion steps into a truth-finding machine, which tells us if he's guilty or innocent." Anticipating my question, he explained, "It's a mind-reading device. We cannot read Ganymedean minds unaided. We can read parts of yours because they are totally exposed. Ours are protected by natural brain barriers that can be broken down only with an artifact."

"But that's a blatant violation of privacy!"

"Our equivalent of your constitution says that while we have the right to a presumption of innocence, we also have the obligation to allow this innocence to be independently tested when circumstances justify it."

"You mean, you *force* the suspect to step into the mind-reading machine?"

"Nobody is forced to do anything. It's the suspect's civic duty to go through the process. Innocent people have nothing to fear. As for the guilty, they usually confess before testing begins." Seeing the look of incredulity on my face, JJ added, "You forget that our planetary MICQ is above 60."

"We'll never build a mind-reading machine. Our defense lawyers and civil libertarians won't stand for it. Our firm belief in democracy won't allow it."

"One of Earth's great politicians[8] famously said that the best argument against democracy is a five-minute conversation with the average voter. Democracy is the least harmful system you have at the moment. When the time is right, you'll improve on it."

"Since we haven't got truth-finding machines yet, what do you suggest we should do to make our system, such as it is, work better?"

"Go over to the inquisitorial version, where the judge is entitled to ask his own questions and instruct searches in an earnest effort to get to the truth; allow the disclosure of previous convictions; abolish dismissal of cases on technicalities; disbar the lawyers who knowingly and willfully distort the facts; eliminate the jury system altogether and replace it with, say, three learned judges; or, if you want to keep the jury, make sure the potential jurors think logically and can follow an argument."

"How would you do that?"

"Ask them a trivial question; for instance:

All humans breathe air.
Gormons are not human.
Therefore, gormons do not breathe air: true or false?"

"This won't happen in a million years!"

"And another point," JJ went on. "Some people say that it's better to let ten guilty persons go free than to see one innocent convicted."[9] Obviously, this principle aims to prevent a particularly abhorrent type of miscarriage of justice. But let's see where it takes us if we subject it to logical reasoning. For simplicity, we'll use rape as an example.

---

[8] Sir Winston Churchill: English soldier, statesman, and writer, 1874–1965.
[9] Formulation attributed to Sir William Blackstone, English jurist, 1723–1780.

The number 'ten' is not meant literally; it is there purely to add shape and emphasis.

'Ten' can be replaced by 'a hundred' or 'a thousand'[10] or, indeed, by any positive integer, however large.

The vagaries of the law will always lead to the incarceration of a few innocents.

Therefore, to avoid sending innocents to prison, all rape suspects must be released.

"When extrapolated to other crimes, the argument quickly gets out of hand because it suggests that unsafe convictions can be avoided only through the complete abolition of punishment."

"Are you saying that we should ditch this principle?"

"It depends on what you hold to be more important: the individual or society. If you favor idealistic morality and humanitarianism and think that an individual's rights take precedence over the welfare of society as a whole, then go with the above conclusion, but don't complain when you find yourself living in a world where crime is rampant and the citizens need to make elaborate arrangements for the safety of their home and family. If, on the other hand, you put society above the individual, then you must accept the likelihood of casualties among innocent bystanders. In any war, even the victor suffers losses, and some of them are caused by friendly fire. I cannot give advice on this one: the decision is entirely yours."

"You think your system is better?"

"In our system, justice is done. Period. Yours is only a legal game. It has very little to do with imparting real justice."

"What kind of crimes do your people commit?"

"Once in a very long while, someone violates a mathematical commandment."

"And, if proven, the punishment is...?"

"I already told you: deportation to Callisto. Or the reformatting of the culprit's personality." JJ leaned back in his chair. "Now," he said, "it's time for your stories."

I finished the last of the coffee and pushed the empty cup aside. "OK, here we go. The couple I have today are about lawyers, but, indirectly, they also touch on mathematics."

A defense attorney is cross-examining a police officer during a trial.

"Did you see my client at the scene of the crime?" he asks.

"No, I didn't. Later, however, I saw him with the other six defendants, running a few blocks down the street."

"Were you in your car at the time?"

"Yes."

"You are asking us to believe that while driving on a street teeming with traffic, you identified seven *specific* people in a sea of pedestrians?"

---

[10] This has already been done by various legal experts.

"I am. They were the only ones running."

"How can you say on oath that there were seven of them? Do you have some special ability to count heads faster than any other human being?"

"I didn't count heads. I counted legs and divided by 2."

The attorney bites his lip and swallows hard. "And how can you be sure that my client was one of those persons?"

"He matched the description given to me."

"By whom?"

"By my colleague who arrived first at the scene."

"You trust your colleagues?"

"With my life."

"*That* much?! Tell me, officer, is there a locker room at the police station? Some room where you change into your uniform for your duties and out of it when you are done?"

"Yes."

"And you have your own locker in that room?"

"I do."

"Do you have a lock on its door?"

"Yes, sir."

"Well, then, how is it that you trust your colleagues with your life, yet you find it necessary to lock your locker in a room you share with them?"

The officer smiles. "Ah, you see, sir, our locker room is in the spur linking the police station with the criminal court, and many times we have spotted defense attorneys using this room as a shortcut."

"Well done the officer!" JJ approved. "Trying to be too clever sometimes makes you walk right into it. An attorney should never ask a witness a question to which he doesn't already know the answer."

"And here's the second story," I said.

At the beginning of a criminal trial in a small town, the assistant district attorney calls his first witness, an elderly woman, and, wanting to avoid later objections from the defense, asks her, "Ma'am, do you know who I am?"

"Of course I know who you are," the woman replies. "I've known you for a very long time. I was your math teacher, remember? In the ninth grade, you pulled out the flowers from the school garden and put a couple in a colleague's bag to have him wrongly accused. In the tenth grade, you shot a neighbor's dog because it had barked after 10 pm on three separate occasions. In the eleventh grade, you worshipped your pocket calculator. And now you think you've become someone important, but you're still the same wicked busybody who derives great pleasure from wrecking other people's lives."

The man, left speechless for a second, recovers quickly and asks, "Do you also know the defense attorney?"

The woman nods. "Very well indeed. He's no better than you. All through high school he behaved illogically, lying and cheating at every opportunity. Today, he's doing the same to his wife, partners, and clients. He's an absolute disgrace to the legal profession."

At this point the judge motions both lawyers to approach and, in very soft tones, says to them, "Should either of you ask the woman if she knows me, I'll throw you in jail for contempt!"

"Not many witnesses can speak so freely and truthfully about your legal practitioners." JJ put his hands on the table. "Thanks for the stories," he said. "I've really enjoyed our little chat. If all goes well, we'll talk some more next time." Then, unceremoniously, he sprang to his feet and promptly walked out of the bar.

Notes After the Meeting

Here are the answers to JJ's questions together with some brief personal comments.

Solution to Question 1. The job applicant V may have used the method of proof by contradiction, or *reductio ad absurdum*,[11] to identify the judge and the defense attorneys among the panelists D, H, and Q. Below is a reasoning chain that solves the problem by this method.

(i) Suppose that Q is a PL.

(ii) Then the opposite of what Q said is true, namely, at least two of D, H, Q (meaning exactly two, or all three of them) are TTs.

(iii) From the assumption that Q is a PL, we conclude that there are exactly two TTs: D and H.

(iv) Since both D and H claim to be the judge, this is impossible; hence, at least one of them must be a PL.

(v) Assumption (i) about the nature of Q has led us to a contradiction of the deduction in (iii), so the assumption is false; therefore, Q is a TT.

(vi) Q's statement about there being at most one (that is, exactly one or none) TT among the three of them is true.

(vii) Q is a TT, so there is exactly one TT in the room with V: Q himself.

(viii) D and H are PLs.

(ix) D and H are the defense attorneys and Q is the judge.

The method of proof by contradiction is a very elegant technique for establishing the truth of a statement when constructive procedures seem unavailable. It consists in assuming the opposite of the statement we want to prove and using this assumption to build up a contradiction. Logic then compels us to conclude that our assumption is false and that, consequently, the original statement must be true.

---

[11] Reduction to the absurd. (Latin)

This technique is not to be confused with the so-called method of proof by confidence, occasionally seen in homework, where the 'handler' assumes the statement of a theorem to be true, tries—and fails—to find a counterexample, and then concludes that the statement must indeed be true. Mathematically speaking, this is a lot of nonsense.

As for V's inference that (ix) follows from (viii) above, well, he must have been persuaded by what JJ's files say on the subject.

Solution to Question 2. The error comes from the misinterpretation of (18.4). Using the basic formula

$$\sin\alpha - \sin\beta = 2\sin\left[\tfrac{1}{2}(\alpha - \beta)\right]\cos\left[\tfrac{1}{2}(\alpha + \beta)\right]$$

with
$$\alpha = \angle ABC + \tfrac{1}{2}\angle BAC,$$
$$\beta = \angle ACB + \tfrac{1}{2}\angle BAC,$$

we reduce (18.4) to

$$\sin\left[\tfrac{1}{2}(\angle ABC - \angle ACB)\right]\cos\left[\tfrac{1}{2}(\angle ABC + \angle ACB + \angle BAC)\right] = 0. \quad (18.8)$$

Since $\angle ABC$, $\angle ACB$, and $\angle BAC$ are the angles of a triangle, they satisfy

$$\angle ABC + \angle ACB + \angle BAC = \pi,$$

which means that the second factor in (18.8) is $\cos\left(\tfrac{1}{2}\pi\right) = 0$, so (18.8) is automatically satisfied and tells us nothing. Therefore, we cannot conclude that triangle ABC is isosceles.

Solvers of trigonometry problems sometimes wrongly assume that

$$\sin\alpha = \sin\beta \quad \Leftrightarrow \quad \alpha = \beta.$$

It is sad but true that in today's schools there is little feeling or liking for geometry and trigonometry.

A Word to the Wise

Trigonometry

Applied scientists and engineers normally measure angles in degrees. Mathematicians prefer to measure them in radians. When asked why, they blame calculus for it. And they are correct: calculus needs radians to yield simple and elegant formulas for trigonometric functions. The connection between the two units is given by the equalities

$$1 \text{ radian} = \frac{180}{\pi} \text{ degrees} \quad \Leftrightarrow \quad 1 \text{ degree} = \frac{\pi}{180} \text{ radian}.$$

The six fundamental trigonometric functions are
$$\sin x, \ \cos x, \ \tan x = \frac{\sin x}{\cos x}, \ \cot x = \frac{1}{\tan x}, \ \sec x = \frac{1}{\cos x}, \ \csc x = \frac{1}{\sin x}.$$

 DO

(i) Here is a list of the most frequently used trigonometric formulas:
$$\sin^2 \alpha + \cos^2 \alpha = 1, \quad \tan^2 \alpha + 1 = \sec^2 \alpha,$$
$$\sin(\alpha + \beta) = \sin \alpha \cos \beta + \cos \alpha \sin \beta, \quad \cos(\alpha + \beta) = \cos \alpha \cos \beta - \sin \alpha \sin \beta,$$
$$\sin^2 \alpha = \tfrac{1}{2}[1 - \cos(2\alpha)], \quad \cos^2 \alpha = \tfrac{1}{2}[1 + \cos(2\alpha)],$$
$$\sin \alpha \cos \beta = \tfrac{1}{2}[\sin(\alpha + \beta) + \sin(\alpha - \beta)],$$
$$\sin \alpha \sin \beta = \tfrac{1}{2}[\cos(\alpha - \beta) - \cos(\alpha + \beta)],$$
$$\cos \alpha \cos \beta = \tfrac{1}{2}[\cos(\alpha + \beta) + \cos(\alpha - \beta)].$$

The last two groups are useful in bringing some trigonometric functions to a form that facilitates their integration.

(ii) Simple trigonometric equations:
$$\sin x = 0 \ \Leftrightarrow \ x = \ldots, -2\pi, -\pi, 0, \pi, 2\pi, \ldots$$
$$= n\pi, \ n = 0, \pm 1, \pm 2, \ldots,$$
$$\cos x = 0 \ \Leftrightarrow \ x = \ldots, -\tfrac{3}{2}\pi, -\tfrac{1}{2}\pi, \tfrac{1}{2}\pi, \tfrac{3}{2}\pi, \ldots$$
$$= \tfrac{1}{2}(2n - 1)\pi, \ n = 0, \pm 1, \pm 2, \ldots.$$

☠ DON'T

(i) It ought to be remembered that trigonometric functions do not behave linearly, therefore, innovative 'equalities' like
$$☠ \quad \sin(ax) = a \sin x, \quad \sin(x + y) = \sin x + \sin y$$
are unwelcome. The same goes for
$$☠ \quad \cos(xy) = \cos x \cos y.$$

(ii) This is wrong:
$$☠ \quad \sin x = 0 \ \Rightarrow \ x = 0,$$
whereas this is simply hilarious:
$$☠ \quad \cos x = 0 \ \Rightarrow \ \cos = 0 \text{ and } x = 0.$$

It seems that zero holds such an overwhelming power over some people that they are prepared to go to any length, including the dismemberment of functions into worthless bits, to satisfy the symbol's insatiable appetite for nullification.

SCAM 19

Civil Litigation

> *The first thing we do, let's kill all the lawyers.*
> William Shakespeare[1]
>
> *Potius ignoratio iuris litigiosa est quam scientia.*[2]
> Marcus Tullius Cicero[3]

I met JJ again at a mathematics conference sponsored by Stanford University in Palo Alto, California. Just like the first time, he was sitting alone at a table in a corner of the hotel bar, gazing at me across the room with dark, inscrutable eyes.

As soon as I sat down, he pushed a piece of paper in my direction. "Here," he said laconically. "For your students."

I glanced at the paper and smiled. The text read something like this.

Question 1 (requires basic algebra). *What, if anything, is wrong with the following assertion and proof?*

Theorem. GAWD *is stronger than* B.L.Z. Bub.

Proof. (i) First, we claim that all nonzero real numbers are positive. Let a and b be any two distinct nonzero real numbers and, without loss of generality, suppose that

$$a < b. \tag{19.1}$$

Dividing both sides in (19.1) by ab, we arrive at

$$\frac{1}{b} < \frac{1}{a}, \tag{19.2}$$

which can be written in the equivalent form

$$\frac{1}{b} - \frac{1}{a} < 0,$$

or

$$\frac{a-b}{ab} < 0.$$

By (19.1), the numerator on the left-hand side above is negative. Since the entire fraction is negative, its denominator must be positive; that is,

---

[1] See footnote 1 on p. 1. From *Henry VI*.
[2] It is ignorance of the law rather than knowledge of it that leads to litigation. (Latin)
[3] Roman orator, lawyer, politician, and philosopher, 106–43 BC.

$$ab > 0. \qquad (19.3)$$

This implies that either both a and b are positive or both are negative; in other words, either all nonzero real numbers are positive or they are all negative. Since, by (19.3), the number ab is positive, it follows that all nonzero real numbers are positive.

(ii) We now go back to the assertion of the theorem. Positive numbers are GAWD's creation because they make perfect sense: they can be used to count objects in the real world, to measure time or distance, and so on. Negative numbers, on the other hand, are meaningless in everyday life: they represent nothing practical; they torment many students beyond endurance; they are truly a devilish invention. Since our proof in (i) shows that, in final analysis, negative numbers do not exist, we conclude that GAWD's work beats B.L.Z. Bub's hands down.

Question 2 (requires elementary logic). A card has four (and only four) statements printed on it, as follows:

On this card exactly one statement is false.

On this card exactly two statements are false.

On this card exactly three statements are false.

On this card exactly four statements are false.

Assuming that each statement is either true or false, how many false statements are there on the card?

"Interesting," I said. "Thanks." I folded the paper and put it in my pocket, waiting for JJ to broach, as usual, a subject of his choice. When he didn't, I took the initiative and said, "Last time, you gave our criminal legal system a real drubbing. What about the civil system? In civil cases the facts are not usually in dispute or can be established more easily. It's a matter of how they are weighed and interpreted."

"This is yet another aspect of your society where lack of logic reigns supreme. It is the province of ambulance chasers and greedy litigation lawyers."

"Citizens are entitled to have their grievances heard and resolved in a court of law," I countered.

"Are you sure? In one of our VOTSIT surveys, my colleague Gan came across a woman, Wom, who had sued a cake manufacturer. You'd better listen to her version of events before making radical pronouncements about who is entitled to what."

GAN: Why did you sue?

WOM: Look at me: I'm vastly overweight. For the last 20 years I've eaten nothing but their cakes, which contain unhealthy, fattening stuff. Their product is responsible for my obesity.

GAN: They didn't force you to eat it. You did that of your own free will.

WOM: Free will?! Have you tasted the cakes? They are so delicious, eating them is addictive. I just couldn't stop myself. Those people ruined my life, so it was only fair, I thought, that they should be held to account.

GAN: What did you do?

WOM: I went to a lawyer. A very kind and clever guy, who said that if we lose, I don't pay him, and if we win, then he'll take 30% of the award money. I was fine with that: 70% of something is better than 100% of nothing. He also said that there must be thousands of other people like me, suffering the effects of eating those cakes, and that it would be a good idea to find as many of them as possible, to strengthen our case. He called it a class action.

GAN: How many did he find?

WOM: About 20,000 in all. That was really smart of him. And, as he had predicted, he won the battle for us. Yesterday, the jury made an award of $10 million compensation and a further $10 million in punitive damages. I'll be rich!

GAN: Are you any good at math?

WOM: I can do numbers, if that's what you mean.

GAN: Then let's do a couple of simple sums. The total award is $20 million. The lawyer's contingency fee is 30%; therefore, he'll take a cool $6 million out of the pot. This leaves $14 million for the 20,000 of you plaintiffs, which means that you will be getting an equal share of $700.

WOM (enraged): But that's peanuts!! The fat cat gets $6 million and we are left with only $700 each? How cruel! How dishonest! I'm gonna see a lawyer!

GAN: Good thinking. Just make sure to choose a kind and clever one.

"Gan's final bit of sarcasm was unnecessary," I said.

JJ shrugged. "Sometimes, when we interview unBRITEs, we can't help ourselves. We are only human—well, temporarily."

"Anyway, there aren't many cases like that."

"You'd be surprised. Here is a condensed sample of four more from Gan's files, involving two male and two female silly egregious claimants, Sec."

GAN: Why did you take your employer to court?

SEC: He requires everybody to be at the office no later than 9 am. One morning, having overslept after an all-night party, I drove as fast as I could to get there on time. The lights, however, were not on my side and I had an accident, in which I broke a leg. If my employer had not insisted on strict punctuality, none of that would've happened. I'm now suing him for $3 million.

GAN: What is the basis of your claim against John Doe?

SEC: Recently I burgled his house. When I was vandalizing the upper floor, the guy came home so I had to flee in a hurry. The only way out that I could see was the downpipe outside the window. But the pipe broke under my weight and I fell to the ground, severely injuring my back. This kept me away from housebreaking for almost a year. I've taken John Doe to court for $4 million because his low quality downpipes caused me a lot of pain and loss of income.

GAN: What are you claiming against the local department store?

SEC: Last month I went there to look at some dresses. As I moved between the displays, I tripped over a large purse, fell to the floor, and badly bruised my left side. I'm suing them for causing bodily harm by allowing shoppers to leave purses around unsupervised. I'm seeking $5 million compensation. Had the purse been damaged as well, I would've asked for $8 million.

GAN: Why?

SEC: The purse that brought me down was mine.

GAN: What is your court action about?

SEC: A week ago, while watching a movie in the local theater, I felt an immediate need to go to the restroom. I got out of my seat, ran up the ramp, and hit my head on the wall at its end.

GAN: Why did you run into the wall? Was there no light for you to see where you were going?

SEC: The passage was well-lit, but the restroom door was at the side of the top landing, not right in front of me. I'm suing the theater chain for $6 million for causing injury by putting the door in the wrong place.

"As a consequence of the large amounts won by the plaintiffs in such cases," JJ said, "don't be surprised if one day shoppers are asked to check all purses and bags when they enter a department store, or if you see signs mounted all over the place in cinemas with inscriptions like *This is a wall. Any attempt to pass through it will seriously damage your health.*"

"How would you prevent litigation of this kind?"

"The judges already have the power. They should dismiss more actions that smack of frivolity. They should also fine the attorneys involved for wasting judicial time. That might make lawyers think twice before accepting to appear in court on behalf of greedy simpletons with laughable claims. And then you'd have to deal with the juries, who often don't understand expert evidence and come up with totally unrealistic awards. Not long ago Gan was called to serve on a jury and saw firsthand what some of these twelve persons good and true are up to. I think you'll find it illuminating to hear his exchanges with another juror, Jur, during their deliberations.

GAN: Did you know what the witness was talking about when he said that the vertical through the center of gravity fell outside the base perimeter, and that, to four significant figures, the ratio of the dynamic pressure of the high-velocity air jet on one side of the dividing membrane and the static pressure of the inviscid fluid on the other was less than the critical value of three to one?

JUR: I don't care. The plaintiff fell and broke his arm.

GAN: He fell because he had not read the gauge and stepped beyond the thick red line of the activity zone.

JUR: I don't care. He said that the line was the wrong shade of red, didn't he? It didn't catch his attention properly.

GAN: He also walked with his body bent forward and leaned out too far over the safety barrier.

JUR: I don't care. He walked the way he did because he expected the platform floor to be at an angle, not horizontal.

GAN: Not true. He had crisscrossed that floor for two years and knew every inch of it. His body leaned forward because he was drunk. The blood test at the hospital showed it.

JUR: I don't care. The test was done without his permission, therefore is inadmissible.

GAN: They tested him as part of an established routine to determine the best course of treatment. Since the judge overruled the lawyer's objection, the test results are in. Are you now overruling the judge? What happened to the plaintiff was entirely his fault. He deserves zip.

JUR: I don't care. The poor guy had a cast on his arm for a month and lost his job.

GAN: He wasn't fired. He resigned. And half a year later he found new employment, much better paid than the old one.

JUR: I don't care. This huge international company has a lot of money and must be made to pay for the suffering inflicted on the man and his family. I say we award him $1 million compensation and another $1 million in punitive damages. An accident like this must not happen again.

GAN: Where do those figures come from? The man made $48,000 a year, or $4,000 a month. The hospital bill came to $2,000. The grand total of his monetary loss is $2,000 in health costs, $4,000 for the month he had his arm immobilized, and $24,000 for the six months he was without a job. Adding, say, a generous bonus of $20,000 for the family, we end up with a grand total of $50,000. This is what we should give him if we want to reward his stupidity and greed.

JUR: I don't care. I'm for the little guy. Let's vote him a million bucks and make his day. And let's vote a second million against the company. That's nothing to them. Just a drop in the ocean.

"Did Gan win the argument or not?" I asked.

"Are you joking? Everyone except him went along with the $2 million suggestion. But then a strange thing happened: the judge turned out to be a very sensible woman, a BRITE who reduced the jury's extravagant figure to $50,000. She also commented on the issue of punitive damages, saying that, in her view, if any such sum were awarded in a case, the money should not go to the plaintiff but to a charity—for example, to one that supports medical research. Gan was very pleased."

"I'm sure. Did he perform some telepathic trickery on the judge and slip his figure and thoughts into her mind?"

"Of course not. You know that we aren't allowed to interfere with humans in this way. Verbal persuasion is one thing. Mental tampering, even at a distance, is another. Had he tried the latter, Gan would've been swiftly dispatched to Callisto and made to watch those terminally boring baseball and cricket games, over and over." JJ leaned back in his chair. "Now," he said, "it's time for your stories."

I finished the last of the coffee and pushed the empty cup aside. "OK, here we go. I've got a couple, and they both mention lawyers. First one coming up."

Four friends—a doctor, a lawyer, a gardener, and a university administrator—meet in a bar. After a few drinks, tongues loosen up a bit and the men start arguing about which of their professions is the oldest. Since they cannot agree among themselves, they decide to use an unimpeachable source: the Good Book.

The doctor says, "Well, it's clear now: mine is the oldest profession because the Good Book describes, almost at the beginning, how GAWD took one of First Man's ribs and used it to create First Woman. That required some fine surgery, so there must have been a doctor around to perform it."

The gardener shakes his head vigorously. "No, no. First Man lived in the Garden of Delights, a truly wondrous place. Only a gardener can keep a garden tidy and beautiful. So my profession is older."

"I'm afraid you are both wrong," the lawyer says. "A lot earlier than the Garden of Delights is mentioned, the Good Book states that GAWD made order out of chaos. You cannot do that and have everything run smoothly and efficiently according to the laws of nature without a lawyer. Obviously, mine is the oldest profession."

At which point the university administrator interjects, "Pardon me, but who do you think created the chaos?"

"Never a truer word spoken," JJ said.

"And here's the second story."

An engineer dies and arrives at the Pearly Gates. An inexperienced winged apprentice, standing in for the Gatekeeper, looks his name up in the Register and shakes his head. "I'm terribly sorry," he chirps, "but you're an engineer; you must go down to the Other Place."

The engineer has no choice and is admitted to Hell. Not long after, he realizes that Hell could do with some improvements, so he starts designing and building a few items. Pretty soon, there is a full air conditioning system, escalators, indoor plumbing, and other comforts, which make the man a popular figure with the underworld hierarchy.

About a year later, GAWD calls B.L.Z. Bub and asks, derisively, "How's the old place doing? Still full of heat, smoke, and misery, as usual?"

The Master of Mischief says, "You may find it hard to believe, but that's all a thing of the past. We've got this clever engineer now, who's worked wonders for us. I'm willing to bet that some day soon my place will be a hell of a lot better—no pun intended—than your boring diaphanous estate."

GAWD's voice hardens. "What are you saying? You've got an engineer down there? He must've been sent to you in error. Turn him over to me at once!"

B.L.Z. Bub laughs. "Absolutely not. He's our man. In fact, I think I'll promote him to Honorary Devil. It's about time engineers' merits were justly recognized."

GAWD is deeply perturbed and angered. "You know you can't keep him. If he's not sent back here right away, I'll sue the pitchfork off your tarred paws."

Amused by GAWD's threat, B.L.Z. Bub laughs even louder. "Oh, yeah? And where do you think you're going to find a lawyer?"

"Good question. As far as attorneys are concerned, that B.L.Z. Bub character has certainly cornered the market." JJ put his hands on the table. "Thanks for the stories," he said. "I've really enjoyed our little chat. If all goes well, we'll talk some more next time." Then, unceremoniously, he sprang to his feet and promptly walked out of the bar.

Notes After the Meeting

Here are the answers to JJ's questions together with some brief personal comments.

Solution to Question 1. If both sides of an inequality are divided by the same (nonzero) number, the direction of the inequality is preserved if and only if the number in question is positive. In our case, (19.2) follows from (19.1) if and only if $ab > 0$, or, equivalently, if and only if a and b are both positive or both negative. When one of them is positive and the other is negative, the argument fails. For example, (19.1) holds with $a = -2$ and $b = 3$ because $-2 < 3$, but (19.2) does not: $\frac{1}{3} \not< -\frac{1}{2}$.

The error is easily spotted by people with a basic knowledge of algebra. A lot harder to spot is the much more subtle incongruity in the 'proof', namely, the use of the concept of negative number as a tool in showing that such a number doesn't exist. In logic—therefore, also in mathematics—this type of reasoning belongs to the class of so-called vicious circle arguments.

Solution to QUESTION 2. We split the reasoning process into simple steps.

(i) Because of the word 'exactly', the four statements are mutually exclusive, so at most one of them can be true; that is, we may have no true statement or exactly one true statement.

(ii) Consequently, at least three statements must be false; that is, we have exactly three false statements or exactly four false statements.

(iii) This means that either the third statement is true or the fourth statement is true.

(iv) Hence, there is a true statement.

(v) By (i), there is exactly one true statement.

(vi) By (ii), there are exactly three false statements on the card.

Although this is a simple problem of logic, not many people seem to have the patience to think carefully through it and find the answer.

A Word to the Wise

Inequalities

 DO

Algebraic inequalities need special care. Here are some of their properties:

$$a < b \Leftrightarrow b > a;$$
$$a < b \Leftrightarrow a - b < 0;$$
$$a < b \Leftrightarrow a + c < b + c;$$
$$a < b, \ c < d \Rightarrow a + c < b + d;$$
$$a < b, \ c > 0 \Rightarrow ac < bc;$$
$$a < b, \ c < 0 \Rightarrow ac > bc;$$
$$a < b, \ a, b > 0 \Rightarrow \frac{1}{a} > \frac{1}{b}.$$

☠ **DON'T**

And here are two very common errors (compare with the correct versions above):

$$☠ \quad a < b \ \Leftrightarrow \ ac < bc;$$
$$☠ \quad a < b \ \Leftrightarrow \ \frac{1}{a} > \frac{1}{b}.$$

☺ **DO**

When solving inequalities, sometimes it is necessary to combine both sides. The following example illustrates the procedure: for $x \neq 0, 1$,

$$\frac{1}{x} < \frac{1}{x-1}$$
$$\Leftrightarrow \ \frac{1}{x} - \frac{1}{x-1} < 0$$
$$\Leftrightarrow \ \frac{(x-1) - x}{x(x-1)} = \frac{-1}{x(x-1)} < 0$$
$$\Leftrightarrow \ \frac{1}{x(x-1)} > 0$$
$$\Leftrightarrow \ x(x-1) > 0$$
$$\Leftrightarrow \ x < 0, \ x - 1 < 0 \ \text{or} \ x > 0, \ x - 1 > 0$$
$$\Leftrightarrow \ x < 0 \ \text{or} \ x > 1.$$

☠ **DON'T**

This handling is completely wrong (compare with the example above):

$$\frac{1}{x} < \frac{1}{x-1}$$
$$☠ \ \Leftrightarrow \ x > x - 1$$
$$\Leftrightarrow \ 0 > -1$$
$$\Leftrightarrow \ \text{the inequality holds for any real number } x.$$

SCAM 20

Statistics and Probability

> *There are three kinds of lies: lies, damn lies, and statistics.*
> Benjamin Disraeli[1]
>
> *Probable impossibilities are to be preferred to improbable possibilities.*
> Aristotle[2]

I met JJ again at a mathematics conference sponsored by Tufts University in Medford, Massachusetts. Just like the first time, he was sitting alone at a table in a corner of the hotel bar, gazing at me across the room with dark, inscrutable eyes.

As soon as I sat down, he pushed a piece of paper in my direction. "Here," he said laconically. "For your students."

I glanced at the paper and smiled. The text read something like this.

Question 1 (requires basic probability). *In the Friendly Shell Game, a man (the Inept Crook) stands in front of a table on which three hollow hemispheres (shells) are arranged in a row, curved side up. One of the shells covers a little bead, while the other two have nothing under them. The Inept Crook shuffles the shells very quickly several times, dragging the bead along the table and covering it with various shells at various times. Then he stops and invites you to guess which shell is hiding the bead, betting, say, $10 that you are not able to guess correctly. Suppose that you accept the bet and, confused by the high speed shuffle, make a random choice and point to one of the shells. The Inept Crook, who knows perfectly well where the bead is, lifts another shell, shows that the bead is not under it, and asks if you want to stick with your original choice or change it. What should you do to improve your chances of winning the bet?*

Question 2 (requires basic probability). *A student walks to her local subway station every morning. The trains stopping at that station run either east to the university or west to the mall. The trains going east arrive at 10-minute intervals, and so do the trains going west. The student walks to the station*

---

[1] English statesman and author, 1804–1881. Some wrongly attribute this quote to Mark Twain (see footnote on p. ix). In fact, Mark Twain *reported* Disraeli's words.

[2] Greek philosopher, 384–322 BC.

at no pre-selected time (that is, her time of arrival is random) and takes the first train that comes, regardless of the direction in which it is going. During a semester, she ends up at the mall four times more than at the university. How is this possible?

"Interesting," I said. "Thanks." I folded the paper and put it in my pocket, waiting for JJ to broach, as usual, a subject of his choice. He needed no special invitation.

"Is it raining?" he asked abruptly.

"I don't know. I haven't been outside since morning."

"The forecast said that today there would be a 30% chance of rain. The meteorologists are infallible."

"Not true. They get it wrong many times," I replied.

"They did in the past, but not any more. Now they cover all bases. Their new probabilistic language is perfect, and, although heavily laden with illogic, may sound quite logical to the mathematically untrained. How can you argue with something like 'Tomorrow there will be a 30% chance of rain'? If it rains, they'll say, 'We told you there was a 30% chance that it would happen; well, it did happen.' If it doesn't, they'll say, 'We told you there was *only* a 30% chance that it would happen; well, it didn't.' You notice that the value of the percentage itself is irrelevant: any could be used (except 0 or 100), and the argument would still hold. It's ironic, though, that this clever use of probability may lead to many meteorologists—especially those employed by the media—getting their walking papers."

"But if they are fired, who will do the measurements, the computations, the interpreting, and the warning?"

"There's no need for that. All you really want is a generic bulletin to be broadcast every day from now until the end of time. I'm thinking of something simple, along the following lines:

Tomorrow there will be a 50% chance of precipitation, a 50% chance that the wind will be strong, and a 50% chance that the temperature will be high.

"A beauty, isn't it?" JJ enthused. "The correctness of the forecast, guaranteed by the use of probability, is further enhanced by the fuzziness of the terms 'precipitation', 'strong', and 'high'."

"Why 50% and not another number?"

"Because some of the world's unBRITEs believe that all probabilities are 50%: either a thing will happen or it won't. Very much in line with my proposed text."

"Droll," I smiled. "But seriously, people often need quantification as to how likely it is that a certain event will happen or not."

"This is fine, so long as they are clear that the number refers to something which *might* happen, and not to something that *will* happen. You have no idea how many times I hear grumbles about 'those crummy meteorologists who said that there was a 95% chance of rain and we didn't see one single drop'. And by the way, I liked your unintentional pun."

"What did I say?"

"You said, 'how likely it is that a certain event will happen or not'."

He was right: in probability terms, a 'certain' event always happens. I should have used the word 'specific' instead.

"Statistics and probability live in a strange land," JJ said, "where good mathematical analysis is performed on data that are not always fully reliable. If you don't observe their fundamental principles, you may 'prove' or 'disprove' anything you like. Statisticians acknowledge that mathematics is important, but some don't believe it to be *the most important* of all sciences. This is in direct violation of the Ninth Mathematical Commandment, which lowers their MICQ to at most 44. During one of his court forays, Gan recorded an interesting case on the VOTSIT, where probability played a decisive role. The manager, Man, of a local branch of a large national corporation had sued his bosses for unfair dismissal. He and the corporation's attorney, Att, appeared before judge Jud to litigate. Here is the gist of the case."

JUD: Why was Mr. Man dismissed?

ATT: We audited his branch, your honor, and found out that 40% of employee absences due to illness were on Mondays and Fridays. This high percentage indicated to us that Mr. Man was not in full control of his workforce. He allowed far too many people to get away with convenient extensions of their weekends.

JUD (to Man): Mr. Att's argument doesn't seem to be without merit. You may have been dismissed for good cause.

MAN: I beg to differ, your honor. There are five working days in a week, so the chance that someone falls ill on any one of them is one in five, which is 20%. The chance that someone falls ill on one of *two* different days is 40%. Therefore, the audit result was to be expected. Mr. Att's argument, on the other hand, shows clearly that he doesn't understand elementary probability at all. It is he who should be fired, not I.

"Did the judge find in favor of the plaintiff?" I asked.

"He did. And Att *was* fired."

"Yeah, right!"

"I also know of a case where an enterprizing young mathematician, Eym, applied probability very ingeniously to get even with a dishonest, unscrupulous bookmaker, Dub, who had refused to pay him a $5,000-win after a horse race because Eym had presented the winning ticket 'too late'."

"This promises to be fun. What 'weapon' did he use?"

"The birthday paradox," JJ said.

"But that's not really a paradox; it's just a problem with an unexpected answer. I'm curious to see how Eym handled it."

"A few days after the race he went back to the bookie's office. Gan, who happened to be there documenting human gambling habits and activities, recorded the scene on his VOTSIT. Have a look."

EYM: I'd like to bet that, of the first 50 customers entering through the door, at least 2 will have the same birthday. What odds can you give me?

DUB (figuring on a calculator): Five to one.

EYM (handing over the money): In that case, I'll risk $1,000.

DUB (about an hour later, paying $6,000 to Eym): By the look of you, sir, you were pretty sure you'd win, weren't you? I think you had me there. So I'll give you another $500 if you tell me where I went wrong.

EYM (accepts the offer): How did you work out the odds?

DUB: Like this:

(i) The probability that any given person P has any given birthday is $1/365$.

(ii) The probability that any other person has a different birthday than P's is $(365-1)/365 = 364/365$.

(iii) The probability that 50 persons have different birthdays is $(364/365)^{50}$.

(iv) Therefore, by the complementary event formula, the probability that at least 2 out of 50 persons have the same birthday is $p = 1 - (364/365)^{50} \cong 0.128$. This yields $(1-p)/p \cong 6.8$, which, to diminish my possible loss, I cut down to 5 to 1.

EYM: You computed the probability that at least another one of those 50 persons shares the same birthday with a *specific* member of the group. You didn't allow for all possible birthday pairings. Here's the correct argument:

(i) The probability that the first person's birthday is any day is $365/365$.

(ii) The probability that the second person's birthday is different than the first person's is $(365-1)/365$.

(iii) The probability that the third person's birthday is different from the preceding two persons' is $(365-2)/365$.

(iv) After exhausting the whole list, the probability that all 50 persons have different birthdays is

$$\frac{365}{365} \times \frac{365-1}{365} \times \frac{365-2}{365} \times \cdots \times \frac{365-49}{365}$$
$$= \frac{365 \times 364 \times 363 \times \times \cdots \times 316}{365^{50}} \cong 0.03.$$

(v) Therefore, by the complementary event formula, the probability that at least 2 persons out of the given 50 share the same birthday is $p \cong 1 - 0.03 = 0.97$, which yields $(1-p)/p \cong 0.03$, or 3 to 100. Had I placed my $1,000 at these odds, you would've paid me the original $1,000 stake and only an extra $30. You should be thankful I didn't bet a much larger sum.

"This is a simplified model," I noted, "where the leap year extra day of February 29, the existence of twins, and other such details are disregarded. But, for practical purposes, it serves as a very good approximation."

"Indeed. If people are intelligent and receptive, they can easily absorb the basics of statistics and probability. Last year Gan held a seminar for mathematically gifted high school students, during which he asked them a few questions to see if they could tell why statistics sometimes screw up the numbers and yield false conclusions. It's interesting to hear Stu's answers."

GAN: The manufacturer of brand A of cars ran an ad saying that out of 1,000 drivers surveyed 943 declared that the next car they would buy would be brand A. The ad then proclaimed that brand A was the best on the market. Was the conclusion correct? If not, why not?

STU: The conclusion is suspect. Since the manufacturer obviously wants to boost the sales of brand A, it is highly likely that the sample of 1,000 drivers surveyed must have been picked from among the brand A owners. Of them, 57 had probably become dissatisfied and wanted a change of car.

GAN: In another ad, the same manufacturer actually states explicitly that it had asked 1,000 *randomly picked* drivers which car they would prefer to own: brand A or the hugely popular brand B? The ad says that 967 of those drivers went for brand A, which means that brand A is the favorite market choice. Is the conclusion reliable?

STU: No, it's not. Brand A is a very good but very expensive car. Brand B is also very good, though not as sophisticated, and costs a lot less. Of course people would like to own the more prestigious brand A. But many of them cannot afford it and will never buy it. The survey question was wrong. It should've asked which car they would actually buy, not which one they would prefer to own.

GAN: The survey ship sent to Earth by an alien civilization reports that the dominant life form on this planet is the penguin. Why would they say that?

STU: It looks as if their probe made just one landing, in Antarctica. You can't call that a proper survey. The sample was too restricted.

GAN: Within an hour of each other, two pollsters asked the same people if they were going to vote for party A or party B. The results differed by 10%. Were either of these results a true indication of the voting intentions of the electorate?

STU: I think not. The number of people interviewed must have been quite small. For example, in a sample of 20 it takes only one person to change his or her mind to induce a 10% swing. In a more respectable sample of 1,000, a swing of that size requires 50 people to change sides. I doubt that 50 people would do so within an hour. The electorate isn't that fickle.

"Gan must've had a very astute audience in his seminar," I said.

"There are lots of promising, young, budding BRITEs around. If only they were all enrolled in one school and serviced by first class teachers...." JJ leaned back in his chair. "Now," he said, "it's time for your stories."

I finished the last of the coffee and pushed the empty cup aside. "OK, here we go. Through sheer coincidence, today's three are on statistics and probability themes. First one coming up."

A physicist, an engineer, and a statistician go on a trip to shoot ducks. Early in the morning, they arrive at the spot, hide in the reeds, grab their guns, and blow the duck whistle. A lone duck darts out nearby, trying to flee the scene.

The physicist takes aim and fires, but the shot goes one meter above the duck. Then the engineer jumps into action, but his shot also misses, passing one meter below the flying bird.

At this moment, the statistician throws his gun into the mud and jumps up, shouting with glee, "Great shooting, guys! You hit it!"

"It's refreshing to see engineers being given a breather," JJ said.

"And here's the second one."

A pollster stops people in the street and asks what they think about a certain issue. At the main gate to the local university campus she approaches a student, who agrees to answer.

"Suppose," the interviewer says, "that you have only one hour left to live. How would you choose to spend it?"

"That's easy," the student says. "I'd like to be in a statistics lecture."

"Why??" the interviewer asks, nonplussed.

The student smiles and says, "Because it would seem to me to last forever."

"I know exactly what the student meant," JJ remarked. "Well, thanks—"

"Wait a minute," I cut him off. "I said that today I had three stories. However, before I tell you the third one, I need to know how well you are acquainted with the map of eastern Europe. Please don't feel offended, but as a Jovian you may not have had the opportunity—"

It was JJ's turn to cut me off. I had the impression that he was slightly irked. "My friend, you forget that I didn't come through your public school system. Which country?"

"Romania."

"Capital: Bucharest. Population: around 22 million. Official language: Romanian.[3] Area: a little smaller than that of Oregon. Main river: the Danube. Main mountain chain: the Carpathians (part of which is sometimes called the Transylvanian Alps). Climate: temperate. Counterclockwise from the southeast, borders on: the Black Sea, Ukraine, Moldova, Ukraine again, Hungary, Serbia, and Bulgaria. Historical figure best known—for entirely the wrong reason—outside the country: Count Dracula.[4] Anything else?"

"Nope," I said, "that's ample. Now here's the story."

On the last day of an international conference on weather forecasting, four participants—an American, a Frenchman, a Bulgarian, and a Romanian—talk around a table during a coffee break.

"Listen," the American says, "now that we've presented our learned papers, off the record and just between the four of us, let's tell the truth: how many times do we *really* get our forecasts right?"

The others look at each other, then the Frenchman shrugs and says, "Okay. But since you started this, you go first."

---

[3] A member of the Romance family of languages rooted in Latin, which also includes Italian, French, Spanish, Portuguese, and Romansh.

[4] Vlad III, Prince of Wallachia (the southern province of present-day Romania), 1460–1476. Better known to Romanians as Vlad Țepeș (pronounced *Tse*-pesh and loosely translated as the Impaler), he inspired the name of the vampire in the novel *Dracula* by Bram Stoker (Irish writer, 1847–1912).

"Sure," the American says. And he explains, "You see, we have many satellites, so many, in fact, that we've lost track of their number. Quite a few of them, placed in geosynchronous orbits above the US, incessantly beam down meteorological data to our powerful computers, which, after almost instantaneous processing, come up with a forecast that turns out to be correct 60% of the time."

The other three are awed: in a notoriously imprecise business like weather forecasting, where everything may, and usually does, change at a moment's notice, a success rate of 60% is something one can only dream of.

The Frenchman goes next. "Well," he says, "we also have satellites. Not as many as the Americans, but our country is much smaller. We also have first class computers, which, when linked to the satellite transmissions, produce a forecast that turns out to be true 50% of the time."

Again, the others are impressed: a half-and-half outcome is truly excellent.

The Bulgarian says, "We don't have satellites. Instead, we have weather balloons. We pack them up with various instruments and let them rise into the stratosphere, from where they radio down a good deal of data. Our computers are bought from the Americans and the French, but we can't afford the very latest models, so we get slightly older machines. Nevertheless, they do a pretty decent job and make weather predictions that are correct 40% of the time."

The others nod their heads appreciatively: here are professional colleagues who, through sheer dedication and skill, manage to overcome their technological shortfalls and come up with a creditable rate of success.

Now all eyes turn to the Romanian, who is keeping quiet. "Come on," they say, "it's your turn. How often is your forecast correct?"

"I'm not telling you," the Romanian says sheepishly. "If I did, you'll just roll about on the floor, choking with laughter."

"We'll do no such thing," the others reassure him. "We're off the record and there's no reason to feel embarrassed, whatever your success rate is."

With great reluctance, the Romanian then says, "Very well. If you really want to know, we get our forecasts right 60% of the time."

A second later, the others start rolling about on the floor, choking with laughter.

"You see? I told you," the Romanian says reproachfully.

"But how can we help it?" the Frenchman says when at last he gets his breath back. "You have no satellites, maybe not even weather balloons for all I know, and your computations are very likely done on pocket calculators. How can you claim the same success rate as the Americans? What voodoo technique are you using?"

"It's not voodoo," the Romanian explains. "It's mathematics. We listen every morning to the Bulgarian forecast and then give ours the other way around."

"Technology isn't always everything. There are times when the laws of probability come in very handy. As for that Romanian fellow, he who laughs last, laughs longest." JJ put his hands on the table. "Thanks for the stories," he said. "I've really enjoyed our little chat. If all goes well, we'll talk some more next time." Then, unceremoniously, he sprang to his feet and promptly walked out of the bar.

Notes After the Meeting

Here are the answers to JJ's questions together with some brief personal comments.

Solution to Question 1. Let us label the shells 1, 2, and 3. Since the bead can be under any one of the shells, we have the following three possibilities, called (a), (b), and (c):

	1	2	3
(a)	bead	–	–
(b)	–	bead	–
(c)	–	–	bead

Whichever shell you choose, your chance of guessing correctly is 1/3. Without loss of generality, suppose that you choose shell 2. A simple analysis now indicates that you have a better winning chance if you change your choice.

(i) If the true position of the bead is as in version (a), that is, the bead is under shell 1, then the Inept Crook will lift shell 3 to show that the bead is not there (he will not lift shell 2 because that was *your* choice); hence, changing your choice (from shell 2 to shell 1) would win you the bet.

(ii) If the true position is as in (b), that is, the bead is under shell 2, then the Inept Crook will lift either shell 1 or shell 3 to show that the bead is not there; changing your choice from 2 to 3 or 1 (depending on which shell had been lifted), you would lose.

(iii) If the true position is as in (c), that is, the bead is under shell 3, then the Inept Crook will lift shell 1 to show that the bead is not there; changing your choice from 2 to 3 would win you the bet.

This analysis leads to the conclusion that, should you change your choice after the Inept Crook shows you one of the shells without the bead, your winning chances increase from the original 1/3 to 2/3.

There is no such thing as a Friendly Shell Game. There is, however, a Shell Game, in which the dealer—a true crook—does not lift any of the three shells to show you where the bead is not. On the contrary, to improve *his* chances of winning from 2/3 to 1 (that is, to make sure he gets your money no matter which shell you choose), he also palms off the bead in the quick shuffle so that the bead is under none of the shells. In our version we called the dealer the Inept Crook because, playing the game by the rules, he will eventually lose his money.

The best thing is not to gamble under any circumstances. Chances are that if you do, you will come across a real crook who will soon make you part with your cash. And these people do not take credit cards.

Solution to Question 2. Clearly, the trains going to the university and those going to the mall do not arrive at the station at the same time. If they did, then the student, who, like all the students I have ever known and taught, is very conscientious and does not want to miss a single minute of class, would take the university train every morning. This means that the trains arrive alternately: one to the university (U), the next one to the mall (M), the one after that again to the university (U), and so on. The student ends up at the mall, in the long run, four times more often than she ends up at the university because the time interval UM (between the arrival of a train going to the university and that of the next train, which goes to the mall) is four times longer than the time interval MU (between the arrival of a train going to the mall and that of the next train, which goes to the university).

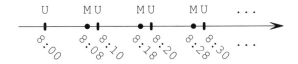

Fig. 20.1.

To see this more simply, let us assume—without loss of generality—that the trains to the university arrive at 8.00, 8.10, 8.20, ... (see Fig. 20.1). Then the trains going to the mall arrive at 8.08, 8.18, 8.28, ... Clearly,

$$UM = 8, \quad MU = 2 \quad \Rightarrow \quad \frac{UM}{MU} = \frac{8}{2} = 4.$$

Since the student goes to the station randomly, she is four times more likely to arrive there during an eight-minute UM sequence (when the next arriving train is an M-train, taking her to the mall) than during a two-minute MU sequence (when the next arriving train is a U-train, taking her to the university).

I want to emphasize that this is a purely mathematical question and that nothing derogatory should be inferred from it about the student's time- and/or money-spending preferences. That students are more inclined to go to the mall than to attend classes at the university is a blatant lie, put out by older people who refuse to understand the high moral values and commendable aims of the younger generation.

A Word to the Wise

Statistics and Probability

The (arithmetic) *mean* of n numbers $a_1, a_2, \ldots, a_n$ is

$$m = \frac{1}{n}(a_1 + a_2 + \cdots + a_n).$$

Popularly, this is known as the *average* of the given numbers. If the numbers are positive, then their *geometric mean* m_g and *harmonic mean* m_h are defined, respectively, by

$$m_g = (a_1 a_2 \ldots a_n)^{1/n}, \quad \frac{n}{m_h} = \frac{1}{a_1} + \frac{1}{a_2} + \cdots + \frac{1}{a_n}.$$

The *median* of these numbers is the one among them that sits in the middle when they are arranged in increasing order (allowing for repeats). This definition can be generalized to include the situations where the number of numbers is even or where the median-value is repeated.

The *mode* is the number occurring most frequently in the set. It may not be unique (that is, there may be more than one such number), and its definition becomes more complicated when the numbers are distinct.

Consider, for example, the set of numbers 1, 4, 5, 6, 6. Here, the mean is

$$m = \tfrac{1}{5}(1 + 4 + 5 + 6 + 6) = 4.4,$$

the median is 5, and the mode is 6.

To get more information about a data set, a quantity called *standard deviation*, denoted by σ, is also computed, by means of the formula

$$\sigma = \left\{ \frac{1}{n}\left[(a_1 - m)^2 + (a_2 - m)^2 + \cdots + (a_n - m)^2\right] \right\}^{1/2}.$$

The standard deviation is a measure of statistical dispersion, showing how spread out the data are from their mean. For example, the above set yields

$$\sigma = \left\{ \tfrac{1}{5}\left[(1 - 4.4)^2 + (4 - 4.4)^2 + (5 - 4.4)^2 + (6 - 4.4)^2 + (6 - 4.4)^2\right] \right\}^{1/2}$$

$$= \sqrt{3.44} \cong 1.855.$$

The *probability* that a certain event will occur is defined as the ratio of the number of favorable cases to that of all possible cases.

☺ DO

The number of cases must be counted very carefully. For example, suppose that a pair of differently colored dice are rolled on a table. The probability $p(6)$ that the sum of the numbers on the dice when they come to rest is 6 is computed as follows:

number of favorable cases: 5 $(1 + 5, 2 + 4, 3 + 3, 4 + 2, 5 + 1)$;

number of all possible cases: 36 (6×6);

therefore,

$$p(6) = \frac{5}{36}.$$

☠ DON'T

Someone inexperienced might wrongly believe the number of favorable cases to be 3 $(1+5, 2+4, 3+3)$, disregarding the fact that each of the combinations $1+5$ and $2+4$ can occur in two distinct ways. Then the probability is incorrectly computed as

 $p(6) = \dfrac{3}{36} = \dfrac{1}{12}.$

SCAM 21

Academic Politics

> *When a true genius appears in the world you may know him by this sign: that all the dunces are in confederacy against him.*
> Jonathan Swift[1]

> *Que dans tous les temps la médiocrité ait dominé, cela est indubitable; mais qu'elle règne plus que jamais, qu'elle devienne absolument triomphante et encombrante, c'est ce qui est aussi vrai qu'affligeant.*[2]
> Charles Baudelaire[3]

I met JJ again at a mathematics conference sponsored by the United States Naval Academy in Annapolis, Maryland. Just like the first time, he was sitting alone at a table in a corner of the hotel bar, gazing at me across the room with dark, inscrutable eyes.

As soon as I sat down, he pushed a piece of paper in my direction. "Here," he said laconically. "For your students."

I glanced at the paper and smiled. The text read something like this.

Question 1 (requires basic geometry). *A sports hall shaped like a rectangular box is 30 m long, 12 m deep, and 12 m high. On one of the two square ends, halfway along the wall and 1 m up from the floor, is an electric socket S (see Fig. 21.1). On the opposite wall, halfway along the wall and 1 m down from the ceiling, is a light bulb B.*

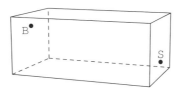

Fig. 21.1.

---

[1] Irish author, journalist, and cleric, 1667–1745.

[2] That mediocrity has always dominated, that is undeniable; but that it reigns more than ever, that it is becoming absolutely triumphant and obstructive, this is what is as true as it is distressing. (French)

[3] French poet and critic, 1821–1867.

An electrician must run a cable from the socket to the bulb in such a way that the cable is at all points in contact with a solid surface (wall, ceiling, or floor) and does not hang in the air. The electrician has only 40 m of cable. Assuming that this is a geometric point-to-point connection and disregarding cable wastage at both ends, can he do the job?

Question 2 (requires basic calculus). *What, if anything, is wrong with the following assertion and proof?*

Theorem. *All students fail all their exams and graduate summa cum laude.*[4]

Proof. (i) First, we show that $1 = -1$. Consider the well-known integral

$$\int \frac{dx}{x} = \ln x. \tag{21.1}$$

We compute the same integral in a different way, by means of the substitution $x = -t$. Then $dx = -dt$ and

$$\int \frac{dx}{x} = \int \frac{-dt}{-t} = \int \frac{dt}{t} = \ln t = \ln(-x); \tag{21.2}$$

hence, from (21.1) and (21.2) it follows that $\ln x = \ln(-x)$. Exponentiating on both sides, we obtain $x = -x$, which, after simplification through x, yields

$$1 = -1. \tag{21.3}$$

(ii) We are now ready to prove the assertion of the theorem. Adding 1 to both sides of (21.3), we find that $2 = 0$, which, on division by 2, becomes $1 = 0$. Multiplying this equality by any positive integer n, we see that

$$n = 0. \tag{21.4}$$

Consider an arbitrary college student, pick any of her courses, and suppose that she has scored a final average of $n\%$ in it. In view of (21.4), the student has failed miserably in this course, so she gets 0 points toward her GPA. But, again by (21.4), we have $0 = 4$, so we deduce that, in fact, her GPA score increases by 4 points. Since both the student and the course were chosen arbitrarily, we conclude that all students fail all their exams and graduate with a GPA of 4.0, which means *summa cum laude*.

"Interesting," I said. "Thanks." I folded the paper and put it in my pocket, waiting for JJ to broach, as usual, a subject of his choice. He needed no special invitation.

"It's amazing," he started abruptly, "to see that academics are ready—and often even eager—to stab each other in the back. I credited them with more sense."

"Why are you surprised? Under their educational veneer, they are just human beings, like everyone else."

---

[4] With highest praise. (Latin)

JJ passed a hand over his slick dark hair. "One expects better from the most learned and cultured representatives of a civilization. They are your intellectual elite, yet sometimes their actions can be utterly illogical. Listen to this conversation between Gan and Pah, professor of ancient history and provost in Gan's university.

PAH: Dr. Gan, one of the proposed dissertation committee members for your PhD student Stu is Dr. Mec from Mechanical Engineering. Is that a wise choice?

GAN: Of course it is. What's wrong with Dr. Mec?

PAH: I'm concerned that an engineer might not know enough about the esoteric doctoral-level mathematics that Stu is working on.

GAN: He knows more than enough. You, on the other hand, know nothing of it, so please don't question my decision.

PAH: But Dr. Mec is a friend of yours. You eat at the same lunch table in the faculty club every day. His inclusion in the committee might be construed by others as making it deliberately easy for Stu.

GAN: *Others?* What *others?* Who's really behind this silliness?

PAH: I just want to ensure that the doctorates we award are above suspicion in external eyes.

GAN: Then leave Stu's case alone and do something about the Center for Research in Artificial Psychology, where they give doctorates without a public defense of dissertations. That should be a worthy cause for you: persuading the employment market not to associate the value of the Center's degree with the acronym of its name. As for Stu, well, since you believe that sharing a lunch table casts doubts on my judgment and Dr. Mec's integrity, I suggest that we put the thumbscrews on the young man and get Dr. Gef, the greatest expert in the field, to come from the other side of the country and examine him.

PAH: Have you ever eaten at the same table with this Dr. Gef?

GAN (astonished, recovers quickly): Nope. Never. But we had a couple of beers together at a conference about ten years ago.

PAH: That's too bad. He's out.

GAN: What about flying Dr. Hae over from Europe? He's the highest authority on Earth in Stu's research area. With him on board—wait a minute—gee, he's no good either: he and I breathed the same air when we met at an international colloquium last summer. Applying your logic, Dr. Pah, we've run out of acceptable candidates. I think we have no choice but to expel Stu from the university. We don't need all that many doctoral graduates anyway.

"Pah was just an inflexible interpreter of rules," I offered. "It happens."

"I also have a record that shows how democracy works in some academic circles. The cast of characters includes the director of the School of Languages, Dir, the five tenured faculty of the English department, generically called Eng, and the building janitor, Jan. One of the Eng guys was, in fact, Gan, who VOTSITed the meeting.

DIR: We have made an offer to an external candidate for the Tallman Chair of Sanskrit.[5] The candidate has written back and asked that his appointment be made jointly in the departments of Rare Languages and English. I have circulated his curriculum vitae to all of you so you could see the high calibre of the man.

ENG: His undergraduate degree is in Latin and his PhD in Old Persian. How does this qualify him to be a professor of English?

DIR: He writes and speaks English fluently.

ENG: So do all the music professors. Yet none of them was given a joint appointment in English. (Raises his hand.) I vote against. Anyone else with me?

DIR (sees the other four hands go up): I don't think you understand the issue here. The Tallman Foundation for Sanskrit Studies has endowed this chair on condition that it be given to a person who stands at least two meters without shoes. Our candidate is 2.11 m.[6] He is a great catch.

ENG: You asked us our opinion and we gave it to you. I'm not going to waste any more time on this. (Leaves the room.)

DIR (to the remaining four faculty): I suggest we vote again.

ENG (raising his hand): I'm still opposed. (Looks at his colleagues.) How about you?

DIR (sees the other three hands go up): You don't seem to understand—

ENG: We understand completely. You are determined to make this appointment regardless of what we think. Why don't you just do it? (Storms out of the room.)

DIR (to the remaining three): Our university badly wants the foundation's money, and hiring such a tall person would be quite a feather in our school's cap. Let's vote once more.

ENG (raising his hand): You told us nothing to make me change my mind. I still vote against. (Looks at the other two, whose hands also go up.)

DIR: This is not a good way forward. Not for me, not for you. Next time you come asking for a new position—

ENG (indignant): Nice try, Dr. Dir. But coercion isn't going to work. (Slams the door on his way out.)

DIR: I must impress upon you that such an important move needs careful consideration. We must not rush the decision. What do you say we take another vote?

ENG (raises his hand, followed by the remaining colleague): Against.

DIR: If we don't appoint this candidate, he'll be going to another institution. There aren't all that many very tall academics in his field. And we'll lose not only the foundation's money, but a lot of face in the community as well.

ENG: I, for one, want to get home tonight. (Walks out the door, just as Jan enters.)

DIR (to Jan): What is it? We are having a meeting here.

JAN: It's very late, sir. I need to clean the room.

DIR (to Jan): Give us a few more minutes. (to Eng) Now, just between the two of us: you scratch my back, and I'll scratch yours. Let's go at it one final time.

---

[5] An old Indic language, still used by some Indian religions.

[6] $2\,\text{m} \simeq 6'7''$; $2.11\,\text{m} \simeq 6'11''$.

ENG: I'm not changing my vote. I'm against the joint appointment.
DIR: Anyone in favor?
JAN (raises his hand): Dr. Dir—
DIR (pointing to Jan): Thank you. Since we have one 'yes' and one 'no', I will exercise my tie-breaking prerogative and vote 'yes'. We are adjourned. (to Jan) The room is all yours.

"Power play and benign malfunctioning," I said.

"Agreed. But academics can also savage one another, with regrettable consequences. You may want to hear what a sample of college professors, referred to as Col, had to say in Gan's recent VOTSIT survey on the subject."

GAN: When you act as an anonymous referee for submissions to journals, do you ever reject articles for the wrong reasons?
COL: Naturally. Sometimes I advise rejection because either the article is not in my research field, therefore it cannot be of interest to anyone, or it is in my field but doesn't quote my own work, or it's too long and I don't have the time or inclination to read it.
GAN: Do you give bad references to colleagues who are up for promotion in other universities? If so, why?
COL: Many times. Usually because the guy is not as good as I am, or is better than I am; or because his wife is prettier, smarter, and more successful than mine; or because I personally dislike him; or just because I have the power to do it.
GAN: Suppose that a newly invented piece of equipment could identify the best of all the applicants for an open position in your department. Would you use it?
COL: Are you for real? This is the last thing I need. I don't want the best candidate. I want to hire someone mediocre like me. A better person would show me up and implicitly be a threat to my career.

"That troublesome word, mediocrity," JJ said. "If you think about it, mediocrity is like a weed: it doesn't amount to much, but it proliferates uncontrollably until it overruns the place."

"This type of response is not typical. Generally, in academia we tend to hire the best applicants."

"Perhaps. Yet such exceptions are encountered more frequently than you think. Stemming from cronyism, bigotry, professional jealousy, nationalism, and other assorted human frailties, they tend to be especially visible in internal departmental promotions. I'm afraid that, in spite of their education and intelligence, not all college professors are BRITES. Some of them, in particular, guilty of covert acts of discrimination, do not rate a MICQ higher than 14 on our scale. Again, see what happened to Gan some years ago when he asked to be considered for elevation to the rank of full professor at his university. He was in competition with a colleague, Lur, a lackluster unproductive researcher, for the only promoted post available that year. Gan was interviewed by a panel consisting of the wishy-washy university president, Wup, the clueless, incompetent dean, Cid, the head of the department, Hod, and the sympathetic director of human resources, Sym. Unbeknown to the others, Gan had his VOTSIT switched on inside his briefcase and recorded the entire conversation.

WUP (to Gan): Why should we promote you, Dr. Gan, instead of Dr. Lur?

GAN: I have published 53 books and 1,345 articles in prestigious journals, as opposed to Dr. Lur's 2 books and 14 articles. I have presented papers at 623 international conferences, as opposed to Dr. Lur's 3. And I have given 428 invited seminars all over the world, as opposed to Dr. Lur's 5.

SYM (whispers to Hod): That's awesome, don't you think?

HOD (to Gan, ignoring Sym): How many articles have you published in our local journal?

GAN: None.

CID: None?!

HOD (whispers to Wup): No support for our efforts to disseminate the subject worldwide. *All* of Dr. Lur's papers are in our in-house journal.

WUP (whispers to Hod): Is that important?

HOD (whispers to Wup): Absolutely crucial.

SYM: Tell us about any special qualities you may have, Dr. Gan, ones that others don't. Anything that makes you stand out.

GAN: Like what?

HOD: For example, can you walk on the ceiling?

CID: Yes, on the ceiling. And on the walls.

WUP (whispers): Really, Dr. Hod...

GAN (takes off his shoes and walks up the wall to his right, then across the ceiling, then down the opposite wall): Did you also ask Dr. Lur to do this?

HOD: We are not interviewing Dr. Lur here; we are interviewing *you*.

CID: None of your business what Dr. Lur can or cannot do.

SYM: How many languages do you speak?

GAN: Several.

HOD: We need proof. Say something in those languages.

GAN: What do you want me to say?

HOD: For example, 'cheers!'

CID: Say 'cheers!'

GAN: Very well. In German: *prost!* In Dutch: *proost!* In Italian: *salute!* In Spanish: *¡salud!* In French: *à votre santé!* In Finnish: *kippis!* In Danish: *skål!* In Russian: *na zdorov'ya!* In Hungarian: *egészségére!* In Turkish: *şerefe!* In Chinese: *gan bey!*

HOD (sarcastically): Is that all?

GAN: In Klingon: *Qapla'!*

WUP (whispers to Hod): This is an impressive number of languages.

HOD (whispers to Wup): You can't really count German and Dutch as distinct, or Italian and Spanish: you noticed how close they sound.

CID: German and Dutch and Italian and Spanish are the same. And Klingon.

WUP (whispers to Hod): I think we've heard enough.

HOD (whispers back): One more question, Dr. Wup, if I may.

WUP (whispers to Hod): All right, but be brief.

HOD (nods to Wup, then turns to Gan): Dr. Gan, walking on the ceiling is not essential to being a full professor. We are more interested whether or not you can cross the lake outside on foot.

GAN (levitates silently above his chair, leans forward 30 degrees, flies through the open window to the lake shore, lands nimbly on his feet, walks across the lake, then shouts back from the other side): Will this do?

SYM (mouth agape): I'll be—

CID: —doggone!

WUP: I think Dr. Gan is truly exceptional.

HOD: Sir, I'm inclined to disagree. You may remember that someone else walked on water before, only to end up nailed to a cross. It's not a feat to be emulated, is it? I'd say Dr. Gan is quite irresponsible. Did you see how he flew through the window without asking permission? He has no respect for authority either. And the word he translated in all those languages!

SYM: But it was *you* who gave it to him—

HOD (energetically): So what? He could've refused and chosen another one instead. I'm clear in my mind that Dr. Gan is an unsavory and subversive character, as well as a wino. By contrast, Dr. Lur, our own homegrown, blue-eyed boy, is exceedingly helpful and doing more around the department than anybody else. He goes out regularly and preaches the gospel of our subject to the people, is well liked by everyone, and, in addition, speaks with a nice local accent.

CID: He knows a lot of words.

WUP (weakly): But his publication record—

HOD: Dr. Lur is a serious thinker. He doesn't rush to put every scrap of thought into a book or article. He lets his ideas gestate. When he goes to print, his results will be Earth-shattering. And look at the references: Lur has three glowing letters whereas Gan doesn't even have a complete set.

SYM: Dr. Lur's letters are from his parents, his next-door neighbor, and his hairstylist. Dr. Gan also has two excellent ones.

HOD: Not three, though.

CID: Only two is not good enough.

SYM: The third hasn't arrived yet.

WUP (to Sym): Who's supposed to write it?

SYM (checks the file): Umm... GAWD.

WUP (to Hod): You asked for a reference from GAWD??

HOD (to Wup): If Gan thinks he's so good, he shouldn't be afraid of GAWD's judgment of his work. The fact that GAWD isn't answering speaks volumes. Everything considered, sir, I have no doubt that Dr. Lur is the man for us. He's the one we should promote.

WUP (lamely): Dr. Gan will be very unhappy. We can't just tell him that we prefer Dr. Lur.

HOD: Let's wait until Dr. Gan goes abroad again and then do it.

"So?" I asked, amused.

"So, a month later, while Gan was receiving the title of Doctor Honoris Causa[7] from a university in another country, Lur got the full professorship. Believe me, there is a lot of dirty play in academia over issues that are really unimportant and should not lead to fighting. But then again, as Henry Kissinger[8] observed, university politics are vicious precisely because the stakes are so small." JJ leaned back in his chair. "Now," he said, "it's time for your stories."

I finished the last of the coffee and pushed the empty cup aside. "OK, here we go. The first one refers to a crucial stage in the career of a would-be academic."

On a warm, sunny day a rabbit is sitting on a tree stump by its burrow in the forest, busily typing on its laptop. A fox comes by and looks at the rabbit.

"What's that you are typing?" the fox asks.

"My PhD thesis."

"On what subject?"

"How rabbits eat foxes."

The fox bursts into uproarious laughter. "Are you crazy? Rabbits don't eat foxes! It's the other way around. What kind of stupid sources are you working with?"

"Come into my burrow and I'll show you," the rabbit replies calmly.

They go inside, then, a few moments later, the rabbit emerges alone and resumes its typing.

After a while, a wolf happens by and stops to look.

"What are you doing there, rabbit?"

"My PhD thesis".

"What's this thesis about?"

"How rabbits eat wolves."

"Do they now!" the wolf bellows heartily. "What kind of nonsensical material have you been reading?"

"It isn't nonsensical. Come into my burrow and see."

The wolf follows the rabbit inside, then the rabbit comes back out by itself and starts to type.

An hour or so later, a bear ambles through the bushes and sees the industrious little rabbit hard at it.

"Hey, rabbit!" the bear rumbles. "What are you writing?"

"My PhD thesis."

"You? A PhD thesis? What are you qualified to write a PhD thesis about?"

"How rabbits eat bears."

"You stupid creature!" the bear bursts out angrily. "You mean, how *bears* eat rabbits."

"No," the rabbit says. "How *rabbits* eat *bears*."

---

[7] Honorary Doctorate. (Latin)

[8] German-born American diplomat and statesman, b1923.

"You're out of your mind, you idiot! You must've been reading the wrong books."

"I've been reading the right books," the rabbit insists. "Follow me into my burrow and I'll show them to you."

When the bear gets inside the burrow, he sees a pile of fox bones in one corner, a pile of wolf bones in the opposite corner, and a huge lion in the middle, picking its teeth with a sharp, blood-dripping claw.

The moral of this story is that it is unimportant what subject you choose for your PhD thesis, or what data you are using. The only thing that really matters is whom you have as your thesis advisor.

"Something every graduate student should know," JJ said.

"The second story corroborates your remark about backstabbing among the high priests of erudition."

The new dean is holding a party in his backyard, to which every faculty member has been invited. After lots of food and lots of drink, the dean gathers all his guests around the pool and says, "Thank you for coming. You've all been assuring me that you will support my every decision and help me run the deanly business smoothly and efficiently. But words are cheap, so show me by deed if you really mean what you say. In this pool is a man-eating shark. A true supporter is expected to have the courage to jump in the water and swim across, risking life and limb for his dean. The person who does that can ask me any favor, and I will grant it unconditionally."

Avoiding each other's eyes and contemplating the drinks in their hands, the guests slowly step away from the edge of the pool. Suddenly, there is a loud splash followed by other watery noises. As everyone turns to look, they see their colleague Dr. Col swimming vigorously from one end of the pool toward the other, with a great hungry shark snapping at his thrashing heels. The man beats the shark by a fraction of a second and pulls himself up out of the water.

"Incredible!" the dean exclaims. "Dr. Col, you have shown loyalty beyond words, so what would you like me to do for you in return? Immediate tenure? A 50% raise? A five-year sabbatical on full pay? Choose, and it shall be done."

"No, sir," Col says, shivering in his soaking suit. "Just tell me which one of my colleagues pushed me into the water."

"A bit weird," JJ commented, "but I see what you mean."

"Not half as weird as your confabulation about Gan's denied promotion."

"You think I made it up? Oh, ye of little faith!" JJ put his hands on the table. "Thanks for the stories," he said. "I've really enjoyed our chat. If all goes well, we'll talk some more next time." Then, unceremoniously, he sprang to his feet, rose about 10 cm in the air, turned slightly to his left to avoid my wide-gaping mouth, and promptly floated out of the bar.

Notes After the Meeting

Here are the answers to JJ's questions together with some brief personal comments.

236   SCAM 21

Solution to Question 1. Imagine that the sports hall is just a rectangular box, which, after a few suitable cuts along the edges, can be flattened out in the plane. In Fig. 21.2 we show the box again, with its six sides labeled by numbers from 1 to 6 (the square walls are 1 and 2, the rectangular walls are 3 and 4, the ceiling is 5, and the floor is 6); thicker lines indicate the edges to be cut.

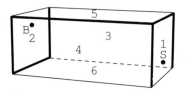

Fig. 21.2.

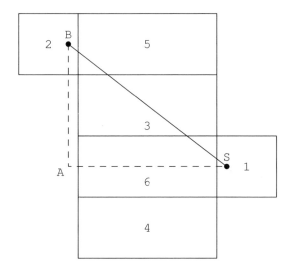

Fig. 21.3.

In Fig. 21.3 we show the flattened box after cutting. Since the shortest distance between two points in a plane region is the length of the straight line segment joining the points (assuming that there are no holes in that region between the two points), the path of the cable in this configuration is along segment SB: the cable runs first on wall 1 from point S to a point on the floor (6), then on the floor to a point on wall 3, then on wall 3 to a point on the ceiling (5), then on the ceiling to a point on wall 2, and, finally, on wall 2 to point B.

To compute the length of this cable, consider the right triangle SAB, where, according to the data,

$$AS = 1 + 30 + 1 = 32 \text{ m}, \quad AB = 6 + 12 + 6 = 24 \text{ m};$$

hence, by Pythagoras's theorem,

$$BS = \sqrt{AS^2 + AB^2} = \sqrt{32^2 + 24^2} = \sqrt{1600} = 40 \text{ m}.$$

The thick line in Fig. 21.4 is the cable path when the box is put back together again.

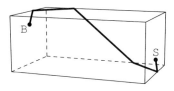

Fig. 21.4.

This is certainly *not* how an electrician would run the cable from S to B. It is much more likely that he would go for one of the 'symmetric' solutions, for example: from S vertically to the ceiling, then along the middle of the ceiling to the opposite wall, and then down that wall to B. As we can verify, such a path is 42 m long. But 2 extra meters are of little significance if what we want is an easy, safe, and practical design.

Mathematical problems, even when seemingly based on real-life situations, are not always meant to offer the most commonsensical or aesthetically pleasing solutions. Like this one, many of them simply aim to teach students how to look at things from a new or unusual angle; in this particular case, literally, how to think outside the box.

Solution to Question 2. Those who paid attention to Calculus II know that the correct result of the indefinite integral in (21.1) is $\ln|x| + C$, where C is an arbitrary constant of integration. Similarly, in (21.2),

$$\ln|t| + C = \ln|-x| + C = \ln|x| + C,$$

which confirms (21.1) and does not generate any mathematical inanity.

My guess is that many students don't spot the error because in the kind of computations they are required to perform, the integral of x^{-1} is generally buried in the middle of a longer chain of operations and the calculator or computer they are using gives them the final result without showing any of the intermediate steps. They could ask to see these steps if they wanted to, but why waste time? After all, the machine knows very well what it is doing, doesn't it?

Sprightly-minded readers may have already worked out other strange assertions that can be 'proved' on the basis of (21.4), such as 'All people in the world are bald', 'Men and women talk in equal amounts', etc.

A Word to the Wise

Limits

Consider an infinite sequence of numbers $a_1, a_2, a_3, \ldots, a_n, \ldots$. If there is a number a such that, going far enough in the sequence, we find that all the remaining terms are close to a to within any prescribed accuracy, then we say that the sequence has the limit a, or tends (converges) to a, and write

$$\lim_{n \to \infty} a_n = a \quad \text{or} \quad a_n \to a \text{ as } n \to \infty.$$

This is an intuitive way of introducing the concept of limit. A fully rigorous definition can be found in any book on advanced calculus.

For example,

$$1, \frac{1}{2}, \frac{1}{3}, \ldots, \frac{1}{n}, \ldots \to 0 \text{ as } n \to \infty,$$

whereas

$$2, \frac{3}{2}, \frac{4}{3}, \ldots, \frac{n+1}{n}, \ldots \to 1 \text{ as } n \to \infty.$$

On the other hand, the sequence

$$1, \frac{1}{2}, 1, \frac{1}{3}, 1, \frac{1}{4}, \ldots, 1, \frac{1}{n}, \ldots$$

has no limit (is divergent) because although its even-numbered terms are steadily approaching 0, all its odd-numbered terms are equal to 1, which means that, however far we go in the sequence, there will always be terms that do not get arbitrarily close to 0.

If $\lim_{n \to \infty} a_n = a$ and $\lim_{n \to \infty} b_n = b$, then

$$\lim_{n \to \infty} (a_n \pm b_n) = a \pm b,$$

$$\lim_{n \to \infty} (ca_n) = ca, \quad c = \text{constant},$$

$$\lim_{n \to \infty} (a_n b_n) = ab;$$

if, in addition, all $b_n \neq 0$ and $b \neq 0$, then

$$\lim_{n \to \infty} \frac{a_n}{b_n} = \frac{a}{b}.$$

The concept of limit can be extended to functions. Thus, we say informally that a function f of a variable x has the limit l as $x \to x_0$ if the values $f(x)$ are arbitrarily close to l when x is sufficiently close to x_0. Obviously, all the points x in the vicinity of x_0 (but not necessarily x_0 itself) must lie in the domain of f so that the numbers $f(x)$ are meaningful.

Further generalizations can be made to (various types of) convergence of sequences of functions and to convergence in so-called metric spaces, where distance is defined in an abstract way.

SCAM 22

Antisocial Behavior

> *A bad neighbor is as great a calamity as a good one is a great advantage.*
>
> Hesiod[1]

> *He is as antisocial as a flea. Clearly, such people are undesirable, and a society in which they can flourish has something wrong with it.*
>
> George Orwell[2]

I met JJ again at a mathematics conference sponsored by Vanderbilt University in Nashville, Tennessee. Just like the first time, he was sitting alone at a table in a corner of the hotel bar, gazing at me across the room with dark, inscrutable eyes.

As soon as I sat down, he pushed a piece of paper in my direction. "Here," he said laconically. "For your students."

I glanced at the paper and smiled. The text read something like this.

Question 1 (requires basic calculus). *A straight railway track FL is 750 m long. Suppose, for simplicity, that the direction from F to L is considered positive. A locomotive is positioned at L facing F, and a mathematical fly is positioned at F facing L. When a signal is given, the locomotive starts running from L toward F at a constant speed of 180 km/h. Simultaneously, the fly starts flying in a straight line from F toward L at a constant speed of 90 km/h and a constant height of 1 m above the track. Soon, at time $t = T$, they collide, after which whatever remains of the fly, stuck to the front of the locomotive, is transported back toward F. Up to the moment of impact the velocity of the fly is positive (it is traveling in the positive direction, toward L); after impact, the velocity of the fly is negative (it is now being carried by the locomotive toward F). Since the switch from positive to negative occurs at time T, it follows that at $t = T$ the velocity of the fly takes the value 0. Consequently, at $t = T$ the fly is instantaneously at rest. But at $t = T$ the fly also becomes attached to the locomotive, so we conclude that the locomotive is also instantaneously at rest at that moment. Hence, the puny fly stops the mighty locomotive dead in its tracks for an instant. How could that happen?*

Question 2 (requires basic algebra). *A card trick starts with the performer*

---

[1] Greek poet, around 700 BC.
[2] See footnote 2 on p. 101.

dealing out a pack of 21 cards. The pack is held upside down and the cards are dealt from the top and arranged face up in a rectangular array with 7 rows and 3 columns, going row by row. The upper half of each card covers the lower half of the card above it in the same column (see Fig. 22.1).

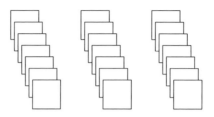

Fig. 22.1.

The dealer asks a spectator to select visually one of the cards in the array and to tell him in which of the three columns the card is located. Then the dealer carefully gathers up the cards in each column (preserving their order in the column by sliding them one under the other) in three subpacks of 7 cards each and puts the subpacks on top of each other, with the one for the column identified by the spectator placed in the middle. This rearranged pack of 21 cards is now turned upside down and the process is repeated three more times. After the last (fourth) array is laid out on the table, the dealer correctly guesses that the card chosen by the spectator is the one at the center of the array (that is, in row 4 and column 2). Why does the chosen card end up there every time, regardless of its position in the initial array?

"Interesting," I said. "Thanks." I folded the paper and put it in my pocket, but instead of waiting for JJ to broach, as usual, a subject of his choice, I asked him, "How will you know that mankind is ready for open contact?"

"I told you," he replied. "Your planetary MICQ needs to be at least 50."

"You misunderstood: my question is not about the test—which, of course, I remember; it's about the testing. Do you monitor the figure at regular intervals?"

"At *very* regular intervals."

"Are you saying that you measure the MICQ of every human on Earth monthly? Or yearly? Isn't this physically impossible?"

"It's not, and we do. You cannot imagine the scale of our technological capabilities."

"When we met in Providence, you told me that Earth's MICQ is now currently somewhere between 25 and 30. How could it be so low? Surely our good people vastly outnumber the bad."

"'Good' and 'bad' are relative terms," JJ said. "Those you call 'good' may not be deemed good enough by the MICQ evaluator. Earlier, we talked about reckless drivers. In the same bottom band, just above hard criminals and unrecoverable drug addicts, our MICQ classification includes another subset of unBRITEs: antisocial individuals. While not exactly breaking the law, or not breaking it in a dramatic manner, these persons nevertheless make a terrible nuisance of themselves by flouting the accepted norms of civilized society and bringing misery to the life of the 'good' citizens. They are profoundly illogical and no better than those who break the First Mathematical Commandment."

"Whom exactly do you have in mind?" I asked, although I knew the answer already.

"Gan did many VOTSIT surveys of JOEs to gauge public opinion on this subject. Here is a particularly interesting one, where the interviewees were college seniors, Sen, with very promising MICQs, who were asked not only to visualize their immediate reaction to the situations described to them, but also to insert a mental arithmetic twist in the 'penalty phase'. Each situation revolves around the behavior of bothersome antisocial types, generically called Bat."

GAN: You are walking or driving in town, or sitting in your backyard or in your house, when Bat passes by and assaults your eardrums with a heavy beat, deafening noise pouring out of his car's open windows. How do you respond?

SEN: The windows go up and click shut. The car speakers burst into a Wagnerian[3] theme at ear-splitting volume. Bat is writhing in his seat, acute pain contorting his face. He tries to turn the system off and fails. The operatic soprano booms up and down her registers, "If you want this to stop, solve the following problem. A customer buys 12 CDs. The store charges $20 for each of the first five, and discounts each of the additional CDs by 25%. How much does the customer have to pay?" "I don't know!" Bat yells in agony, his eyes bulging out of their sockets. "I don't care! I hate math!" "Then enjoy!" the soprano sings tumultuously.

GAN: What should Bat have replied?

SEN: $205. The first five CDs cost $5 \times \$20 = \$100$. Each of the remaining seven costs 25% less, which is $15, giving a subtotal of $7 \times \$15 = \$105$.

GAN: You are in a public place when Bat's cell phone suddenly rings out. He answers it and, without retiring to a private corner, launches into a conversation with the other party, conducted at quite a few decibels more than would be acceptable. Your reaction?

SEN: The phone jumps out of Bat's hand and rams itself down his throat. He tries to yank it out, unsuccessfully, and starts to choke. A voice speaks from the phone: "This torture will end if you can solve the following problem. A cube of side 3 cm is filled with water 12 times and its contents are poured into a larger cube of side 9 cm. How high is the water in the latter?" "3 cm," Bat sputters with great difficulty. "Incorrect," the voice says, then the phone sound changes to slow, lugubrious music.

GAN: Why would Bat say 3 cm?

---

[3] Richard Wagner: German composer, 1813–1883.

SEN: No idea. From what I've seen, when people are wrong about something mathematical, they are usually wrong in a big way. The problem is easy. The volume of the small cube is $3 \times 3 \times 3 \, \text{cm}^3$. The total volume of water (in cubic centimeters) poured into the larger cube is, therefore,

$$3 \times 3 \times 3 \times 12 = 3 \times 3 \times 3 \times 3 \times 4 = 9 \times 9 \times 4.$$

Since the area of any of the large cube's faces is $9 \times 9 \, \text{cm}^2$, the depth of the water is $(9 \times 9 \times 4)/(9 \times 9) = 4 \, \text{cm}$. I didn't multiply out the factors on purpose, to make cancelation easier.

GAN: You are shopping in a large store and notice that Bat, having changed her mind about an item, discards it in the wrong place. What would happen?

SEN: Bat is unable to let go of the item, which seems stuck to her fingers. The store speakers are blaring out at her, "If you want to free your hand, answer the following question. Two coins add up to 35 cents. One of them is not a dime. What are they?" "The question is unfair: it doesn't have an answer," Bat shouts angrily. "Oh, yes, it does," the speakers retort. "Now go through the entire store, collect all randomly abandoned items, and put them in their proper spot on the shelf."

GAN: Clearly, Bat tried to reason. What do you think went wrong with her argument?

SEN: She probably took 'one of them is not a dime' to mean 'neither of them is a dime'. The answer is a quarter and a dime, that is, 25 cents and 10 cents. One of them—the quarter—is not a dime, so the condition in the problem is satisfied. Bat was very short on logic.

GAN: While trying to park at a department store, you see Bat leaving his empty pushcart in the parking place next to his car instead of walking it to the specially designated area nearby. How does this end?

SEN: Bat starts the engine, ready to drive off. The empty pushcart rushes at his car, denting its side and attaching itself to it. Bat curses, gets out and pulls the cart away. The cart waits until Bat is back in the car, then does it again. This happens several times until, livid with fury, Bat asks an attendant to get rid of the cart. "It's impossible," the attendant says. "But the cart will stop attacking if you answer the following question. Out of 120,000 items produced by a factory, 18 are defective. What is the rate of rejects per 100,000 items?" "I don't care! I don't have a calculator!" Bat shouts. "Just make this crazy thing go away!" The attendant shakes his head and says, "I pity your car, sir."

GAN: How should Bat have solved the problem?

SEN: It's a matter of proportionality. If you ignore the four final zeros in 120,000, the ratio of 18 to 12 is 3/2. Doing the same with 100,000, the number whose ratio to 10 is 3/2 is 15. Simple arithmetic doesn't need calculators.

GAN: Early in the morning or late at night, or in daytime during the weekend, as you try to relax in your home, you hear the incessant yapping or barking of a dog left by Bat out in his backyard. How do you react?

SEN: Bat is whisked away to the Country of Bowwows.[4] He is stark naked on a cold and soggy lawn, wandering about on all fours and howling in distress. Inside the house, a dog, propped up on its hind legs at the window, nose wrinkled in a satisfied grin, gestures insultingly with its forepaws and growls, "You'll be spared this ordeal if you tell me the sum of all the positive integers from 1 to 100." "I'll tell you nothing! Choke on a bone and die, you stupid mutt!" Bat spits through his chattering teeth. "Wrong answer," the dog growls back. "I hope you like wet grass."

GAN: People should not be cruel to animals, let alone to other people, including their neighbors. How would you have computed the sum?

SEN: Set 50 and 100 aside, then start adding the pairs of numbers working in from both ends: $1 + 99 = 100$, $2 + 98 = 100, \ldots, 49 + 51 = 100$. There are 49 such pairs that, collectively, add up to 4,900. Throwing in 50 and 100, you get the grand total of 5,050. Since Bat has no insight and no imagination, he should learn to be kind to his pet and respect the peace and quiet of those living nearby.

GAN: You stroll along the street and see Bat waiting for her dog to poop on the pavement or on the grass in front of someone else's house. Bat then leads the animal away without collecting its 'deposit'. What happens next?

SEN: When she arrives home, Bat discovers that her own front lawn is covered with dog poop. A note taped to the door reads, "The lawn will become clean again if you can answer correctly the following question. A student is asked to compute the mean—or average, as is commonly known—of her six test scores. She operates with all the numbers on the list but then, mistakenly believing that there are only five scores, she gives the answer 90. What is the correct value?" Bat is very annoyed and says loudly, "I wish that all math lovers die a gruesome death!" At this point, she turns the paper over and discovers a second message, which reads, "Bad temper will not get you off the hook. Now go to the corner shop and stock up on little plastic bags and scoops."

GAN: How should Bat have handled the problem?

SEN: The average score is the aggregate score divided by 6. She divided by 5 and got 90, so the aggregate score is $90 \times 5 = 450$. In turn, this yields the actual average of $450/6 = 75$, which earns the student a pass. Someone or something has put a double whammy on Bat, because she is bad at both arithmetic and cursing.

GAN: You are in a public restroom and notice that, after using the facility, Bat is walking away without bothering to flush the water...

"Ugh! Gross!" I interrupted. "I think I've heard enough."

"JOEs troubled by antisocial behavior don't mince words, and well brought-up and properly schooled young ones are no exception. Their logic is simple:

Good citizens treat other humans with respect and courtesy.
Selfishness makes one insensitive to other people's needs and rights.

---

[4] Apologies to Jonathan Swift (see footnote 1 on p. 227) and his Country of Houy-hnhnms, where horses rule the land and dominate the brutish race of debased humans called Yahoos. (Yup, this is where that word comes from.)

Bat is very selfish.
Therefore, Bat is not a good citizen.

I found it quite intriguing to see how easily Gan's interviewees, when prompted, associated this type of behavior with lack of basic mathematical knowledge."

I took a sip of coffee from my cup. "I assume that you also surveyed a bunch of wrongdoers?"

"Naturally. We wanted to know what makes them stray from the norm. No surprises there, I have to say. Here is an example collected by a female Gan, who interviewed a typical Bat as she was trying on some dresses in the fitting room of a department store."

GAN: How many dresses have you got here?
BAT: Don't know. Maybe a dozen.
GAN: You've thrown a few to the floor and are trampling them underfoot.
BAT: Those I've tried already.
GAN: Why didn't you put them back on their hangers?
BAT: Not my job.
GAN: Don't you think that you are damaging them?
BAT: Not my dresses.
GAN: Don't you want to leave the fitting room in good order for the next shopper?
BAT: Not my store.
GAN: Have you always been so untidy and inconsiderate?
BAT: No. All this started when I realized that I could get away with it.

JJ steepled his fingers under his chin. "This is a bigger problem than you think, with obvious causes. It appears that some families don't teach children proper manners and respect for their fellow citizens. Equally, there may be schools where such deficiencies are not detected or eliminated. And the blame should also attach to the legal system and other assorted agencies that don't act more decisively to put an end to loutish behavior, particularly when it contravenes the law or is a direct threat to public health and the well-being of society at large. Until mankind learns how to deal with its slobs and boors, its collective MICQ will remain well below the 50 mark."

"You mentioned cell phones," I said. "Did Gan ever have one ring during a lecture?"

"Yes, soon after he got his university position."

"How did he react?"

"You mean," JJ said, "other than frown severely and wait for the thing to be switched off? He told the class his Rule Number 1 about cell phones: that they would not be permitted to ring in his presence unless, the university being a place of culture, they played a classical tune."

"But such tunes—"

"—are available, I know. In fact, Stu tried it on Gan once as a prank. The exchange went like this.

GAN: (hears a cell phone playing Beethoven's *Für Elise*[5]): Whose phone is it?

STU (smiles): Mine, sir. You said that, by your rule, classical tunes were acceptable.

GAN (smiles back benevolently): They are. To be precise, as mathematicians must be, I called that Rule Number 1. Can you name the tune being played, and its composer?

STU (uncomfortable): Err...no, not really.

GAN (continues to smile): When you heard me mention Rule Number 1, you should've guessed that there would at least be a Rule Number 2. Well, there is one, and it says that not recognizing the identity of the tune in question means that the necessary cultural standard has not been met.

STU (tentatively, getting wiser): Is there a Rule Number 3?

GAN: Yes. According to Rule Number 3, I'm the one who makes all the rules around here.

A question formed in my mind. "Have you ever had a Bat-like Ganymedean since your civilization's MICQ went above 50?"

JJ looked sheepish. "Yes," he admitted grudgingly. "We've had a handful of cases. Nothing major, mind you, but, yes, it's happened."

"Let me guess: destination Callisto?"

"Oh, no. We couldn't allow our penal colony to be contaminated by such deeply antisocial behavior."

"So what did you do with them?"

"I'm afraid we brought them to Earth." JJ leaned back in his chair. "Now," he said, "it's time for your stories."

I finished the last of the coffee and pushed the empty cup aside. "OK, here we go. I've got two for you today. First one coming up."

A student is browsing around in a curio shop when a man walks in and says to the owner, "I'm the environmental chief officer in the mayor's office. This town is plagued by rats and my mission is to clean it up. I've tried everything, but nothing works. Today someone suggested that perhaps you may have a solution. If you do, please tell me what it is; I'll pay good money for it."

"As it happens," the shop owner replies, "I do have a solution. Here." He rummages under the counter and brings out a little mechanical rat. "You wind up this thing, put it on the pavement, point it toward the river, and let it go."

"And this will get rid of all the rats in town?" the customer asks incredulously.

"It will. First, you pay me one hundred bucks for the toy."

Clearly desperate and willing to try anything, the mayor's man pays the money, takes the mechanical rat, turns the key in its back, goes out into the street, puts the toy down facing the river, and lets it go. The mechanical rat starts running toward the river, making loud rat noises. Suddenly, from all over the town, thousands and

---

[5] Ludwig van Beethoven: German composer and pianist, c1770–1827. *Für Elise* (For Elise) is the title of his Bagatelle in A minor.

thousands of rats come out of hiding and join the little toy. When it reaches the river, the toy jumps into the water, followed by the live rats, who drown *en masse*.[6]

The student has witnessed the entire event open-mouthed and rubs his eyes to convince himself that he is not dreaming. Then he steps up to the counter with purpose and says, "Sir, here is a hundred bucks. Do you also have a tiny mechanical calculus instructor?"

"Ha!" JJ exclaimed. "What about a tiny mechanical problem student?"

"Now the second story: an unconventional application of 'logical reasoning' to zoology."

Theorem. All alligators have square-shaped bodies.

Proof. We break the argument into three parts.

(i) An alligator is long both on its back and on its belly, but it is green only on its back. Therefore, an alligator is longer than it is green.

Looking at its back, we see that an alligator is green both from snout to tail and from side to side, but that it is wide only from side to side. Consequently, an alligator is greener than it is wide.

Combining these two statements, we conclude that an alligator is longer than it is wide.

(ii) An alligator is wide both on its back and on its belly, but it is green only on its back. Hence, an alligator is wider than it is green.

Looking at its back, we notice that an alligator is green both from snout to tail and from side to side, but that it is long only from snout to tail. Thus, an alligator is greener than it is long.

From these two statements we deduce that an alligator is wider than it is long.

(iii) Consider an arbitrary alligator. By (i), the beast is longer than it is wide. By (ii), it is wider than it is long. Therefore, the alligator is as long as it is wide, which means that its body is square-shaped.

Since the alligator was arbitrarily chosen, it follows that all alligators have square-shaped bodies.

"Would you pass a student who wrote a mini-theory like this for you?" JJ asked.

"I would. For making original use of the language and logical argument."

"So would I." JJ put his hands on the table. "Thanks for the stories," he said. "I've really enjoyed our little chat. If all goes well, we'll talk some more next time." Then, unceremoniously, he sprang to his feet and promptly walked out of the bar.

Notes After the Meeting

Here are the answers to JJ's questions together with some brief personal comments.

---

[6] All together. (French)

Solution to Question 1. *Reasoning grosso modo,*[7] we readily accept that at the moment of impact the locomotive maintains its speed of 180 km/h and never stops. To explain the apparent anomaly in the question's conclusion, however, we need to analyze the data by means of a simple calculus argument. If v_f and v_l are the velocities of the fly and locomotive, respectively, then their speeds are

$$|v_f| = 90 \text{ km/h} = 25 \text{ m/s},$$
$$|v_l| = 180 \text{ km/h} = 50 \text{ m/s}.$$

This means that their relative speed (the rate at which they approach each other) is $25 + 50 = 75$ m/s, so they collide after $750 \div 75 = 10$ s, at a point $10 \times 25 = 250$ m from F (see Fig. 22.2).

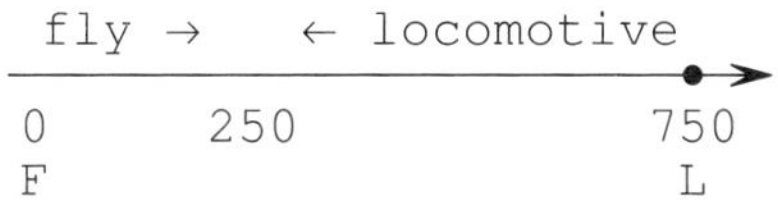

```
   fly →        ← locomotive
 ─────────────────────────────●──→
   0           250            750
   F                           L
```

Fig. 22.2.

Since in rectilinear motion with constant speed we have

$$\text{distance} = \text{speed} \times \text{time},$$

the position $s_f(t)$ of the fly at time $t > 0$, measured from F, is given by the formula

$$s_f(t) = \begin{cases} 25t, & 0 \leq t \leq 10, \\ 750 - 50t, & t > 10. \end{cases}$$

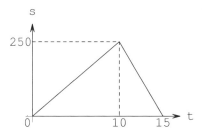

Fig. 22.3.

It is clear that this function is continuous everywhere, including at $t = 10$, where both branches yield the same expected value of 250. The graph of $s_f(t)$ is shown in Fig. 22.3. However, the function is not differentiable at $t = 10$ (the

---

[7] Roughly, approximately. (Latin)

moment of impact), since

$$s'_f(t) = \begin{cases} 25, & t < 10, \\ -50, & t > 10, \end{cases}$$

that is, the two branches yield different values of the derivative to the left and right of $t = 10$. Since $v_f(t) = s'_f(t)$, this means that the velocity of the fly is not defined at the moment of impact: it jumps from 25 to -50 without going through the value 0. Consequently, the fly did not stop the locomotive.

It is wrong to believe that a quantity changing from positive to negative or from negative to positive must necessarily assume the value 0 somewhere in between. Discontinuous functions do occur in real life, and they need proper consideration.

By the way, if you haven't noticed, the height of the traveling fly above the ground is just a red herring. The other numerical data also have no influence on the final answer, as long as they are nonzero.

Solution to Question 2. We will work out the answer in four simple and more or less identical steps.

(i) Suppose that, in the first (initial) array, the chosen card is in row i_1 (unknown to the dealer). Clearly,

$$1 \leq i_1 \leq 7. \tag{22.1}$$

When the cards are gathered up in the manner described in the problem (all face up), the chosen card is in position i_1 from the bottom of its column subpack, therefore in position $i_1 + 7$ from the bottom of the new pack (because another column subpack of 7 cards is placed under its own subpack). When the new 21-card pack is turned upside down, the chosen card is in position $i_1 + 7$ from the top. Adding 7 to each of the three sides in (22.1), we find that

$$8 \leq i_1 + 7 \leq 14,$$

which means that the chosen card can be anywhere between the eighth and fourteenth positions from the top of the new pack.

(ii) In what row does the chosen card end up when the second array is dealt out? Let i_2 be that row. Since the dealing is done row by row, if the chosen card is eighth from the top, it lands in row 3 (rows 1 and 2 are filled by the first 6 cards, so the eighth card is in row 3). If the chosen card is in the other extreme position—that is, fourteenth from the top—it lands in row 5 (rows 1, 2, 3, and 4 are filled by the first 12 cards, so the fourteenth card is in row 5). To sum it up, we have

$$3 \leq i_2 \leq 5. \tag{22.2}$$

Reasoning as in step (i), we see that, when the new pack is formed again and turned upside down, the chosen card is in position $i_2 + 7$ from the top. Adding 7 to each of the three sides in (22.2), we find that

$$10 \leq i_2 + 7 \leq 12,$$

which means that the chosen card can be anywhere between the tenth and twelfth positions from the top of the new pack.

(iii) In what row does the chosen card end up when the third array is dealt out? Let i_3 be that row. Following the argument in (ii), if the chosen card is tenth from the top, it lands in row 4 (rows 1, 2, and 3 are filled by the first 9 cards, so the tenth card is in row 4). If the card is twelfth from the top, it still lands in row 4. Hence, $i_3 = 4$. Reasoning once more as in steps (i) and (ii), we see that, when the last pack is formed and turned upside down, the chosen card is in position $i_3 + 7 = 4 + 7 = 11$ from the top.

(iv) In what row and column does the chosen card end up when the fourth (and final) array is dealt out? The card is eleventh from the top. The first 9 cards fill rows 1, 2, and 3, so the eleventh card will be the second one in row 4, that is, in the center of the (7×3)-array.

The trick can be made to look much more impressive if, after learning which card had been chosen by the spectator, the dealer pretends not to have the answer yet and does some further arbitrary shuffling of the pack before 'guessing' the chosen card.

A Word to the Wise

Differential Calculus

We assume that the functions f and g mentioned below depend on a real variable x and that they are differentiable in their domain. To keep the formulas simple, x is not shown explicitly in them.

☺ DO

(i) Differentiation is a linear operator; that is,

$$(f + g)' = f' + g',$$
$$(cf)' = cf', \quad c = \text{constant}.$$

In particular, note that constant factors can be put in, or taken out of, a function under differentiation.

(ii) Brackets must be used when the function to be differentiated consists of several terms; for example,

$$\frac{d}{dx}(3x^2 - 2x + 4) = 6x - 2.$$

(iii) Products and quotients of functions are differentiated by means of the rules
$$(fg)' = f'g + fg',$$
$$\left(\frac{f}{g}\right)' = \frac{f'g - fg'}{g^2}.$$

(iv) The chain rule of differentiation states that if $f = f(g)$ and $g = g(x)$, then the derivative of the composite function $f(g(x))$ is
$$\frac{d}{dx} f(g(x)) = \frac{df}{dg} \frac{dg}{dx}.$$

☠ **DON'T**

(i) The privilege enjoyed by constants cannot be extended to the active variable; therefore, the likes of

☠ $\quad \dfrac{d}{dx}(x^2 e^x) = x^2 \dfrac{d}{dx}(e^x) = x^2 e^x$

are unacceptable.

(ii) Many times, careless writing carries a heavy price. I have seen the derivative of, say, $3x^2 - 2x + 4$ given as

☠ $\quad \dfrac{d}{dx} 3x^2 - 2x + 4 = 6x - 2x + 4.$

Not inserting parentheses around $3x^2 - 2x + 4$ has led in this case to the differentiation of only the first term, $3x^2$.

(iii) The following 'simplified' formulas are invalid:

☠ $\quad (fg)' = f'g',$

☠ $\quad \left(\dfrac{f}{g}\right)' = \dfrac{f'}{g'}.$

☺ **DO**

Here is a minimal table of derivatives of basic functions:
$$(x^a)' = ax^{a-1}, \quad a = \text{constant} \neq 0;$$
$$(\sin x)' = \cos x;$$
$$(\cos x)' = -\sin x;$$
$$(e^x)' = e^x;$$
$$(\ln x)' = \frac{1}{x}.$$

SCAM 23

Mathematicians Versus Engineers

> *Il faut se méfier des ingénieurs, ça commence par la machine à coudre, ça finit par la bombe atomique.*[1]
>
> Marcel Pagnol[2]

> "I hope you will forgive my dreadful curiosity, but I should like awfully to know—what is your profession?" I replied that I was an engineer. She emitted an involuntary exclamation, and said "Why, I thought you were a gentleman!"
>
> Herbert Hoover[3]

I met JJ again at a mathematics conference sponsored by Washington University in St. Louis, Missouri. Just like the first time, he was sitting alone at a table in a corner of the hotel bar, gazing at me across the room with dark, inscrutable eyes.

As soon as I sat down, he pushed a piece of paper in my direction. "Here," he said laconically. "For your students."

I glanced at the paper and smiled. The text read something like this.

Question 1 (requires basic algebra). *Prove or disprove the following statement:*

$N(n) = n^2 - 81n + 1681$ *is prime for any positive integer* n.

(Recall that a prime number is a positive integer whose only factors are 1 and itself.)

Question 2 (requires basic calculus). *What, if anything, is wrong with the following assertion and proof?*

Theorem. *All engineering mathematics is incorrect.*

Proof. (i) First, we show that $0 = 1$. We write the formula of integration by parts

$$\int f(x)g'(x)\,dx = f(x)g(x) - \int g(x)f'(x)\,dx \tag{23.1}$$

with

$$f(x) = e^{-x}, \quad g(x) = e^x.$$

---

[1] One must mistrust engineers, for they begin with the sewing machine and end up with the atomic bomb. (French)
[2] French film director and playwright, 1895–1974.
[3] The thirty-first president of the United States, 1874–1964.

Then
$$f'(x) = -e^{-x}, \quad g'(x) = e^x,$$
so, by (23.1),
$$\int e^{-x} e^x \, dx = e^{-x} e^x - \int e^x(-e^{-x}) \, dx = e^{-x+x} + \int e^x e^{-x} \, dx$$
$$= e^0 + \int e^{-x} e^x \, dx = 1 + \int e^{-x} e^x \, dx.$$

Subtracting the integral terms on both sides above, we obtain
$$0 = 1. \qquad (23.2)$$

(ii) We return to the assertion of the theorem, and let n be the number of correct results obtained by engineers. Multiplying (23.2) by n, we arrive at $0 = n$, or, equivalently, $n = 0$. This means that all engineering mathematics is wrong, and the theorem is proved.

"Interesting," I said. "Thanks." I folded the paper and put it in my pocket, waiting for JJ to broach, as usual, a subject of his choice. He needed no special invitation.

"Do you ever get into conflict with other users of mathematics?" he asked me abruptly.

"Not often. Normally, they mind their own business and I mind mine. I get more hassle from mathematicians sometimes."

"Of course. In each discipline, many researchers arrogantly tend to believe that only their field is important, and that people working in other fields are wasting their time. The picture across disciplines is more complicated. Personally, I find it fascinating to see what humans with contiguous specialties think of each other's professions. We touched on this subject a couple of times before, when we were not very complimentary about engineers. I think we need to clarify our position a bit better."

"Engineers are fine as long as they don't speak ill of mathematics and don't claim to be as good at it as we are," I said.

"Agreed. In fact, it's a small minority among them that is responsible in part for the adoption of our Tenth Mathematical Commandment."

"If they take liberties with us, then we must go on the offensive and defend our turf. And this time we are playing at home, in a manner of speaking, so fire at will."

JJ relaxed in his chair. "You don't have to tell me any stories today. Instead, I'll let you hear a small sample that Gan collected when he amused himself by doing a VOTSIT job on a mathematician, Mat, and an engineer, Eng, who shared a lunch table with him at a joint scientific conference. Although from the same—very large—university, Mat and Eng didn't know each other. Gan kept quiet and listened to their conversation, which went something like this."

MAT: I'm a mathematician.
ENG: I'm an engineer.
MAT: I know. I've read your name badge. You have my sympathy.
ENG (pretends to be annoyed): You just can't help yourself, can you? Why do mathematicians think that they're a class above us? After all, mathematics doesn't have as many practical uses in real life as engineering. Allow me to convince you of this with a story.

On a nice summer day, two friends, A and B, are taking a trip in a balloon. They are having a great time when, suddenly, the weather changes and a powerful storm sweeps over the area, covering the sky in thick black clouds and buffeting the balloon this way and that. An hour later, when the storm is over and the skies are clear again, the two friends look around and realize that they have completely lost their bearings.

"Not to worry," A says confidently. "If I let out some helium, we'll get sufficiently close to the ground and ask someone for directions."

The maneuver is successful and, as the balloon loses height, A and B spot two men walking in the field below.

"You, down there!" A shouts, hands cupped to his mouth.

When the two men stop and look up, A asks, "Where are we?"

The two men, C and D, exchange telling glances. After a few seconds, C shouts back, "You are in a balloon!"

"Oh, no!" A says to B. "Of all the people in the world, we had to come across a mathematician!"

B is puzzled. "How do you know he's a mathematician?"

"Simple," A explains. "First, he didn't reply immediately; he thought before he answered. Second, his answer is 100% correct. And third, the answer is completely useless."

MAT (laughs slowly): Ha, ha. Why did you cut the story short?
ENG: I didn't. That's the whole of it.
MAT: No, it isn't. This is how it really ends.

After he hears C's answer, A bends over the rim of the basket and shouts back, "Thanks for nothing, buddy!"

Down on the ground, C looks at D and says, "Fancy that: a balloon driven by an engineering manager."

D is intrigued. "How do you know he's an engineering manager?"

"Obvious," C replies. "Before he asked, he was totally lost. He's still lost after I answered him, but now he blames me for it."

MAT: To say that mathematics is useless flies in the face of the evidence, my friend. You must admit that you are relying on a fair bit of math yourself when you design your bridges, skyscrapers, and space rockets. We don't hold ourselves above you guys. The truth is that mathematicians and engineers have a kind of symbiotic relationship. You provide us with challenging mathematical problems arising in your work, and, in return, we give you techniques and solutions you can put to good use. Where we differ is in the details of our methodologies. As practitioners, you don't generally bother so much with rigor, which is fine by us. But you shouldn't claim that rigor is not really necessary, that it's just a 'nicety of form' and therefore a waste of time and effort. We want absolute rigor in our arguments because we want our results to represent absolute truths. Also, we are better than you at generalizing and abstract thinking. In response to your story, let me offer you one of my own, to even things up a bit.

Four colleagues—two mathematicians and two engineers—travel by train to the same conference. At the station, the engineers buy two tickets, while the mathematicians buy only one.

"What's going on here?" one engineer whispers to the other. "How will they get away with a single ticket between the two of them?"

"I don't know," the second engineer replies. "Let's see what they do."

The four board the train and sit together, facing each other, in the middle of a car. Some time after the train's departure the conductor shows up, announces loudly "Tickets, please!" and starts punching the passengers' tickets. Calmly, the mathematicians get up and walk to the opposite end of the car, where they lock themselves in the restroom cubicle.

When the conductor reaches the restroom, he tries the door, and, finding it locked, knocks and says, "Tickets, please!" A ticket slides out under the door. The conductor punches it, then slides it back and walks away into the next car.

The engineers cannot help admiring the mathematicians' ingenuity and resolve to follow their example on the return journey.

At the end of the conference the four meet again in front of the station's ticket office, and this time the engineers buy only one ticket. The mathematicians simply walk by.

"Amazing!" the engineers marvel. "We really must see what they do this time."

On the train, when the conductor arrives, the two engineers do exactly as the mathematicians did when they had traveled down, and lock themselves into the restroom. Once the engineers are inside, the mathematicians wait a good few seconds, then get out of their seats and walk to the end of the car, on their way to the adjacent one. Just before entering the connecting passage, they stop in front of the restroom and one of them knocks on the door, saying in a commanding voice, "Tickets, please!"

ENG: Who's exaggerating now? We've got practical minds. We would never fall for such a stupid trick.

MAT: Nor would we play one on you. We have good manners and freely share our discoveries with everybody. Anyway, it was you who started the mudslinging, not I.

ENG: You realize that this story can also be told with the roles of the engineers and mathematicians reversed.

MAT: It can, but it wouldn't have as much credibility.

ENG: Why not?

MAT: You know very well why. In real life, you engineers get large grants and can afford to travel comfortably. We mathematicians receive hardly any, so we have to do it on the cheap.

ENG (remembers): What niceties of form were you referring to earlier?

MAT: Well, for example, you don't bother to check that a series representation for a function is convergent before using it to compute approximate values. You just start adding up the terms. If the series diverges, then adding more terms makes things worse, not better. Your bridge might collapse.

You don't believe in mathematical induction: you make an assumption, verify it in a number of cases, and conclude that it's valid in all cases. Your skyscraper might topple over.

And you don't care for proofs of existence, uniqueness, and stability of the solution to a model.

ENG: I admit that we compute solutions numerically. What use is it to us to know that there is some strange element in some obscure abstract space, which satisfies the requirements of our model in some ultra-sophisticated fashion? We want numbers. We want graphs. We believe in things we can touch and feel.

MAT: But if you don't know that there *is* a solution, your computations and graphs mean nothing. Your rocket might blow up on the launch pad. You need us to do the background work for you, to show when your results are valid and when they are not.

ENG And does this justify your lofty attitude toward us? If our lines of work are equally important, each in its own way, then why are you so dismissive of our profession?

MAT: We are not!

ENG: Yes, you are. Your haughtiness has even penetrated popular humor, as the next story illustrates.

An engineer ambles along the shore early in the morning, when, suddenly, he stubs his toe on something buried in the sand. Intrigued, he reaches down and picks up what looks like a very ancient bottle. As he loosens the cork, a genie shoots out with a strong rush of air and lands in front of him on the beach.

"Don't tell me," the engineer smiles, "now you'll be granting me three wishes because I set you free."

The genie sighs. "Afraid not. What with inflation and everything, it's going to be only one wish. So think very hard before you tell me what you want."

"I don't need to think at all. I know exactly what I'd like to have more than anything else. If you really are who you claim you are, then show me how to build a time-travel machine."

"Ooohh!" the genie says, rolling his eyes. "I do have great powers, but they

don't extend that far. Time travel involves tearing up the fabric of the space–time continuum, which would drain the universe of all its energy and extinguish it. Your request is beyond unreasonable—ask me something else."

"That's too bad. I really had my heart set on that project." The engineer shrugs and adds, "If you can't fulfill my first-choice wish, then fix it for my profession to be taken seriously and to be respected by other experts, especially by mathematicians."

The genie stares at the engineer for a couple of seconds. "By mathematicians? Oh, well... how big would you like the time-travel machine to be: just for driver, or driver and passenger?

MAT: I was right: you need to shake that chip off your shoulder.

ENG (finishes eating): I don't think we'll ever see eye-to-eye on these issues, so we'd better agree to disagree and leave it at that. I've got a plane to catch.

MAT: So do I. Funny you should mention the plane. Here is a final little story that, I hope, makes peace between our professions once and for all.

Two engineers and a mathematician from the same university are flying back from a conference on scientific computing. The mathematician sits on the aisle, with the engineers beside him, in the middle and window seats. The mathematician has taken off his shoes, to feel more comfortable.

Shortly before landing, the engineer by the window tries to stand up and says, "If you'll excuse me, I'd like to get a soda."

"I'll get it for you," the mathematician says. "I also want one." And he walks down the aisle toward the back of the plane.

While he is away, the engineer grins mischievously, takes out the chewing gum from his mouth, and sticks it in one of the mathematician's shoes.

A couple of minutes later, the mathematician returns with a bottle of soda.

"That's mighty nice of you," the engineer says, taking the bottle and emptying it in one long, thirsty slurp.

"Good idea," the other engineer says. "I think I'll have one, too. But please don't trouble yourself on my account," he tells the mathematician, who was already half out of his seat.

"No trouble at all," the mathematician says. "This gives me a chance to exercise my legs."

With the mathematician off to the galley, the second engineer snickers and does exactly what the first one had done, using the other shoe.

The mathematician comes back and hands a soda bottle to his colleague, who thanks him and downs its contents with relish.

A few moments later the chief flight attendant asks the passengers to bring their seats to the upright position, fasten their seat belts, and prepare for landing. As the mathematician puts his shoes on, he immediately realizes what had happened.

"Come on, guys," he says in a sad, reproachful voice, "how long do we, mathematicians and engineers, have to continue this silly vendetta? This contempt for each other's professions? This unjustifiable animosity? This spitting of gum in shoes? This adding of laxatives to soda bottles?"

ENG (ready to leave the table): Which flight are you on?
MAT: The 2:45.
ENG (with visible relief): I'm on the one after that. See you later.

"Who do you think won the debate?" I asked JJ.

"Hard to tell. They both put forward good arguments. But if I had to make a choice, I'd say that Mat won on points, by a small margin."

"You aren't saying this just to please me, are you?"

"Nope. I thought Mat's stories were better." JJ put his hands on the table. "I've really enjoyed our little chat," he said. "If all goes well, we'll talk some more next time." Then, unceremoniously, he sprang to his feet and promptly walked out of the bar.

Notes After the Meeting

Here are the answers to JJ's questions together with some brief personal comments.

Solution to Question 1. In situations of this nature, where we have to deal with a statement that depends on a generic positive integer n, it is helpful to examine a few simple special cases, to get a feel for what is happening and an idea of what to expect. Here, for example, we have

$$
\begin{aligned}
n = 1 &\Rightarrow N(1) = 1601, \\
n = 2 &\Rightarrow N(2) = 1523, \\
n = 3 &\Rightarrow N(3) = 1447, \\
n = 4 &\Rightarrow N(4) = 1373, \\
n = 5 &\Rightarrow N(5) = 1301.
\end{aligned}
$$

Since the numbers 1601, 1523, 1447, 1373, and 1301 are all prime, we might be tempted to guess that $N(n)$ is prime for all positive integers n. However, verifying just a few special values does not constitute a proof. Indeed, were we to persevere and compute $N(n)$ for $n = 5, 6, \ldots, 80$, we would obtain a prime number every time. However, for $n = 81$ we get

$$N(81) = 81^2 - 81 \times 81 + 1681 = 1681 = 41^2,$$

which, clearly, is not prime. Hence, the statement is false.

In mathematical circles, the procedure that consists in verifying a statement like this for a few specific values of n, finding it to be true for those values, and then proclaiming it to be true for all values of n is facetiously referred to as the Principle of Engineering Induction. It is an unhealthy practice that should be avoided at all costs because it leads to exam failure.

The legitimate counterpart of this is the Principle of Mathematical Induction, which is used sometimes to prove that a statement of the form $S(n)$, whose formulation involves a generic positive integer n, is true for all $n \geq n_0$, where n_0 is given. Such a proof consists of two phases:

(i) the verification phase, when we check that $S(n_0)$ is true;

(ii) the inductive phase, when we assume that $S(k)$ is true for some arbitrary k and, on the basis of this assumption, show that $S(k+1)$ is also true.

If we accomplish (i) and (ii), then we conclude that $S(n)$ is true for all $n \geq n_0$.

Solution to Question 2. *The mathematical error is obvious: in this case we are dealing with indefinite integrals, so formula (23.2) is valid up to an arbitrary constant of integration: a fact that not many students seem to remember, to their peril.*

A Word to the Wise

Integral Calculus

We assume that the one-variable functions f and g below are integrable.

☺ DO

(i) Integration is a linear operator; that is,

$$\int [f(x) + g(x)]\,dx = \int f(x)\,dx + \int g(x)\,dx,$$

$$\int cf(x)\,dx = c \int f(x)\,dx, \quad c = \text{constant}.$$

The second equality shows that, as in differentiation, constant factors can be put into, or taken out of, the integrand.

(ii) Brackets should be used when the function to be integrated consists of several terms; for example,

$$\int (3x^2 - 2x + 4)\,dx = x^3 - x^2 + 4x + C.$$

Notice the presence of the differential dx and the arbitrary constant C of integration.

(iii) If, additionally, f and g have integrable first-order derivatives f' and g', then the following formula of integration by parts is valid:

$$\int f(x)g'(x)\,dx = f(x)g(x) - \int f'(x)g(x)\,dx.$$

DON'T

(i) Consider the 'manipulation'

$$\int \frac{1}{x^2+1}\,dx = \frac{1}{2x}\int \frac{2x}{x^2+1}\,dx = \frac{1}{2x}\ln(x^2+1).$$

There are two errors here: the variable of integration is taken into the integrand as if it were a constant, and the arbitrary additive constant is missing on the right-hand side. While the latter is a common bad error, the former is simply unforgivable.

(ii) Sloppy writing may induce errors like this:

$$\int 3x^2 - 2x + 4 = x^3 - 2x + 4.$$

The integrand has not been enclosed in brackets and the differential dx is missing.

(iii) The following 'simplified' formulas are invalid:

$$\int f(x)g(x)\,dx = \left(\int f(x)\,dx\right)\left(\int g(x)\,dx\right);$$

$$\int \frac{f(x)}{g(x)}\,dx = \frac{\int f(x)\,dx}{\int g(x)\,dx}.$$

(iv) This manner of handling can only be called scandalous:

$$\int \tan x\,dx = \int \frac{\sin x}{\cos x}\,dx$$

$$= \int \frac{\text{in}}{\text{co}}\,dx \qquad\qquad\text{(a,b)}$$

$$= \frac{\text{in}}{\text{co}}\int dx \qquad\qquad\text{(c)}$$

$$= \frac{\text{in}}{\text{co}}x. \qquad\qquad\text{(d)}$$

In case you don't understand why anyone should feel outraged by this, here is the crime list:

(a) x, the variable of the functions sin and cos, was canceled as a factor;

(b) 's' was canceled as if sin and cos were products of s, i, n and c, o, s, respectively;

(c) the new 'integrand' (in)/(co) was taken out of the integral as a constant (independent of x);

(d) the arbitrary constant of integration was omitted.

Before worrying about (c) and (d), noting (a) and (b) is enough to make any reasonable person want to jump off a tall building.

☺ DO

A minimal table of basic indefinite integrals might be useful:

$$\int x^a \, dx = \frac{x^{a+1}}{a+1} + C, \quad a = \text{constant}, \ a \neq -1;$$

$$\int \frac{1}{x} \, dx = \ln|x| + C;$$

$$\int e^x \, dx = e^x + C;$$

$$\int \sin x \, dx = -\cos x + C;$$

$$\int \cos x \, dx = \sin x + C.$$

And here is another integral that comes in handy many times:

$$\int \frac{f'(x)}{f(x)} \, dx = \ln|f(x)| + c,$$

where $f'(x)$ is the derivative of $f(x)$.

SCAM 24

The Evolution of Knowledge

> As we acquire more knowledge, things do not become more comprehensible but more mysterious.
>
> Albert Schweitzer[1]
>
> The beginning of knowledge is the discovery of something we do not understand.
>
> Frank Herbert[2]

I met JJ again at a mathematics conference sponsored by Xavier University in Cincinnati, Ohio. Just like the first time, he was sitting alone at a table in a corner of the hotel bar, gazing at me across the room with dark, inscrutable eyes.

As soon as I sat down, he pushed a piece of paper in my direction. "Here," he said laconically. "For your students."

I glanced at the paper and smiled. The text read something like this.

Question 1 (requires basic calculus and mechanics). *What, if anything, is wrong with the following assertion and proof?*

Theorem. *Physical sciences on Earth are nonsensical.*

Proof. We show this in two steps.

(i) First, we prove that every object on Earth's surface has infinite weight. Consider a small, heavy sphere S at the end of a rigid arm OS of negligible weight that rotates without friction in a vertical plane, around a fixed point O on the ground (see Fig. 24.1).

The arm, initially at rest with S at point $(0, a)$, rotates downwards until the sphere hits the ground at time T at point $(a, 0)$.

Let $x(t)$ and $y(t)$ be the coordinates of S at an arbitrary intermediate moment of time between $t = 0$ (when S starts moving) and $t = T$ (when S stops moving). Then, by Pythagoras's theorem,

$$x^2 + y^2 = a^2 \quad \text{for all } t, \ 0 \leq t \leq T.$$

Using the chain rule to differentiate this equality with respect to t and dividing through by 2, we arrive at

$$xx' + yy' = 0 \quad \text{for all } t, \ 0 \leq t \leq T, \tag{24.1}$$

---

[1] German theologian, musician, philosopher, and physician, 1875–1965.
[2] American writer of science fiction, 1920–1986.

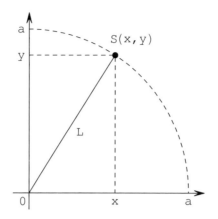

Fig. 24.1.

where $x'(t)$ and $y'(t)$ are the derivatives of $x(t)$ and $y(t)$ with respect to t, which represent the components of the velocity vector of S. From (24.1) we obtain

$$y' = -\frac{xx'}{y} \quad \text{for all } t,\ 0 \leq t \leq T. \tag{24.2}$$

At the moment $t = T$ when S hits the ground we have

$$x = x(T) = a, \quad y = y(T) = 0, \quad x' = x'(T),$$

so, by (24.2),

$$y'(T) = -\infty. \tag{24.3}$$

On the other hand, the second law of motion, as formulated by Newton,[3] states that the mass of S multiplied by its acceleration is equal to the sum of the forces acting on S. Writing this law in vertical projection, we have

$$my'' = -mg,$$

where m is the mass of the sphere, $y'' = d^2y/dt^2$ is the vertical component of its acceleration, and g is the acceleration of gravity. The sign '$-$' occurs because the weight of S, which, according to our assumptions, is the only vertical force acting on S, is directed downwards, that is, in the negative direction of the vertical coordinate axis. We divide both sides by m and rewrite the last equality as

$$y'' = -g.$$

---

[3] Sir Isaac Newton: English physicist, mathematician, astronomer, and natural philosopher, 1643–1727.

Integrating once, we find the vertical velocity of S to be

$$y'(t) = -gt + C_1,$$

where C_1 is an arbitrary constant of integration. Since S is initially at rest, we have $y'(0) = 0$, so $C_1 = 0$; hence,

$$y'(t) = -gt. \qquad (24.4)$$

Integrating a second time, we get

$$y(t) = -\tfrac{1}{2} gt^2 + C_2,$$

where C_2 is another arbitrary constant of integration. Since the initial position of S is characterized by $y(0) = a$, we obtain $C_2 = a$; therefore,

$$y(t) = -\tfrac{1}{2} gt^2 + a. \qquad (24.5)$$

S hits the ground—that is, $y = 0$—at time T; hence, by (24.5),

$$y(T) = -\tfrac{1}{2} gT^2 + a = 0,$$

from which

$$T = \sqrt{\frac{2a}{g}}.$$

We now replace T in (24.4) and find that

$$y'(T) = -gT = -g\sqrt{\frac{2a}{g}} = -\sqrt{2ga}. \qquad (24.6)$$

But, by (24.3), $y'(T) = -\infty$, so, comparing with the last equality, we must have

$$\sqrt{2ga} = \infty.$$

Since a is a finite length, this can happen only if g is infinite. Therefore, the weight of any object, which is equal to the product of its mass and the acceleration of gravity g, is also infinite.

(ii) Turning to the assertion of the theorem, we remark that the centers of black holes are the only places with an infinite gravitational pull in the known universe. By part (i) of the proof, the gravitational pull on the surface of Earth is infinite, so our planet must be at the center of a black hole. And since in these strange locations all physical laws break down, we infer that physical sciences on Earth make no sense whatsoever.

Question 2 (requires basic set theory). *Which contains more points: a line or a circle?*

"Interesting," I said. "Thanks." I folded the paper and put it in my pocket, then 'jumped' on JJ before he could broach, as usual, a subject of his choice.

"I'd like to know, if I may, if string theory is correct," I said. "Does it indeed explain and unify the main forces present in our physical universe, as its proponents claim?"

"I don't know," JJ replied casually.

"What do you mean—you 'don't know'? You're so much more advanced than us. Surely you've figured out a master theory that covers everything?"

"We have, but it's not like yours."

"So string theory is wrong?" I asked, disappointed.

"I'm not saying that at all. Generally, there is more than one way to look at things, interpret them, and fit them together in a coherent picture. Has your theory been verified? Ours has."

"I don't think we have any concrete proof of its validity yet. But then you knew that already. Although a beautiful construct, it's not easily explained to the layman."

"Recently," JJ said, "Gan gave a public lecture for high school students in his city, titled 'The Earth Is Almost Spherical, but the Universe Is Infinite'."

"Strange title, and an unusual use of the adversative *but*."

"Done purely for effect, no doubt. The idea was to try to make the young audience understand how the universe is explored and interpreted by the mind. Here is what he said, and what Stu asked him during his talk. Gan recorded the event on his VOTSIT."

GAN: If we assume that the universe is infinite, then it seems reasonable to expect it to have infinitely many phenomenological manifestations. Intelligent beings, however, with their finite structure in space and time, can have knowledge of only finitely many of them at any given moment. Since it is in their nature to study, explain, classify, and correlate all known phenomena, such beings are always in search of unifying theories that present a clear global picture of everything currently within their knowledge. As soon as a new phenomenon is discovered and catalogued, they go back to the drawing board and start looking for a more general theory than the one before, which places the new phenomenon in context with the old ones and weaves this enhanced knowledge into another plausible and coherent pattern. Let us consider a very simple example. Suppose that the population on some faraway planet operates quantitatively in a different way than us—say, by means of awareness of energy levels and direct absorption of the size of things through a special sense. For a long time, the only natural number they knew was 1, representing the individual, the identity. Then suddenly they discover the existence of a new natural number, 2. With this increased knowledge about the numerical universe, they look for a 'theory' that unites 1 and 2 and in which (i) each number is 'created' by the other number (that is, neither can *generate itself*) but (ii) each number may reproduce through combination with the other number.

STU: Why these particular rules?

GAN: Theories, by their nature, are restrictive in certain respects. When you think about it, rule (i) makes some kind of logical sense since no phenomenon is expected to be its own cause. The reason for (ii) is more obscure and I can't pretend to know it. As you can see, I'm not one of these beings.

STU: Would you like to be one of them?

GAN: No. Not yet. Not for a long time. Maybe some day, when I get bored working with so many numbers.

STU (takes a small calculator out of his pocket): Why don't you use a calculator?

GAN (determined to nip this heckling in the bud): Calculators are for wimps. People with guts do sums in their heads.

STU (slides the calculator back into his pocket, pretending he never took it out): So what theory do their mathematicians invent, then?

GAN: After due consideration, they decide to define a binary operation $*$ on the set $\{1,2\}$ and to construct their 'theory' in the form of a table:

$*$	1	2
1	2	1
2	1	1

The table is short for

$$1*1 = 2, \quad 1*2 = 1, \quad 2*1 = 1, \quad 2*2 = 1.$$

It is obvious that the rules imposed on the 'theory' are satisfied: 1 and 2 do not 'create' themselves since $1*1 \neq 1$ and $2*2 \neq 2$; at the same time, 1 is reproduced through combination with 2, in either order. The operation $*$ is called *commutative* because $1*2 = 2*1$. The table is symmetric with respect to its leading diagonal.

STU: This was not a requirement.

GAN: No, it's pure coincidence. Anyway, some time later, mathematicians in that strange population discover the existence of a new natural number, 3. Immediately they set about constructing a more general 'theory', which must incorporate 3 and tie it to 1 and 2. They also refine the formulation of the rules, asking that (i) no number can generate itself, (ii) each number generates one and only one other number, and (iii) each number may reproduce itself in combination with some of the other numbers. Once again, the new 'theory' is developed in the form of a table for a new binary operation $\circ$ on the set $\{1,2,3\}$:

$\circ$	1	2	3
1	2	1	3
2	1	3	2
3	3	2	1

You can now easily imagine how our hypothetical population will act when they discover 4 and the subsequent naturals.

STU: Is this the only possible table that satisfies their rules for 1,2,3?

GAN: No, it isn't. Here is another one, for a different binary operation ◊ on the same set:

◊	1	2	3
1	3	2	1
2	1	1	3
3	1	3	2

This one also satisfies rules (i)–(iii).

STU: I prefer the first table.

GAN: Why?

STU: It's neater. Each row and each column contains all the numbers 1,2,3 in a different order and the table is symmetric, which makes the operation—what did you call it?—commutative.

GAN: Yes, but in those beings' understanding of science this might not mean that the first 'theory' is easier or truer than the second. Remember that they regard numbers as enigmatic entities and not as the obvious, intuitive concept they are to us. They have a completely alien way of thinking.

STU: What happens when they discover that the natural numbers are infinitely many? What kind of table would they be able to draw for an infinite set?

GAN: By that time they may have progressed from tables to a more sophisticated and abstract way of handling numbers. Knowledge develops and accumulates in many different ways. It may be gained through observation and experiment, or through imaginative thinking. Of course, it always needs testing and refining. Acquiring knowledge may be likened to climbing a pole: the higher you get, the more things to be discovered and investigated come into view. Your horizon broadens continuously, and will go on doing so for as long as you have enough energy to climb. Close to hand, we have the example of Newton's mechanics, which for a long time explained very well what was happening to a moving body in our limited planetary universe. Later, when our universe expanded to include motions at speeds comparable to the speed of light, Newton's mechanics had to be taken to a higher plane, becoming a particular case of what is called the theory of relativity, developed by Einstein.[4] And now scientists are already talking about going even higher than that, to allow for motions at transluminal speeds. All the theories designed to explain the workings of the physical world are nothing but increasingly complex mathematical models.

STU: Given their complexity, how do scientists manage to study these models?

GAN: Good question. The first thing you do with a mathematical model is simplify it. For example, if you look at propagation of heat in a metal rod, make the rod perfectly cylindrical, insulate its lateral surface, and assume that the rod material is homogeneous—in other words, its physical properties are the same at every point. You must be careful, though, not to get carried away and oversimplify. Einstein himself said that one should try to make everything as simple as possible,

---

[4] Albert Einstein: German-born American theoretical physicist, 1879–1955.

but no simpler. Once you solve the simpler model and get a fair idea of what's going on, you can start considering additional features; for example, in the heat problem that I mentioned you may allow the material properties to change from point to point, the cross-section of the rod to be variable, and so on.

STU: That pole of learning is infinitely high, isn't it?

GAN: I can safely say that it is. And there's something else that I'll say, my young BRITE friend: you've got a sharp, analytic mind and I urge you most seriously to pursue the study of the wonderful subject of mathematics when you go to college.

"Good advice," I approved.

"Absolutely. You can never have too many skilled mathematicians." JJ leaned back in his chair. "Now," he said, "it's time for your stories."

I finished the last of the coffee and pushed the empty cup aside. "OK, here we go. Today I've got a couple. The first one illustrates Gan's point about starting with a simple model."

A mathematician, a statistician, and an engineer go to the races and place bets on the horses. After the race, they stop at the bar and order drinks.

"I don't understand how I could lose," the engineer says. "I measured all the horses, computed their mechanical strength and energy, and calculated their maximum speeds on the track. What went wrong, I wonder?"

"I know what you did wrong," the statistician says. "You did not allow for the dependence of performance on weather conditions and quality of the ground. I did a statistical analysis of the runners' results during the last year and put my money on those with the highest winning probability."

"Yeah," the engineer retorts. "If you're so smart, how come you didn't win a single bet?"

At this moment the mathematician takes out his wallet, and the other two see a thick wad of bills neatly arranged inside. Realizing that their friend was a big winner, they immediately ask him what method he used to identify the right horses.

"Well," the mathematician explains, "I did as I normally do in my research. I made a mathematical model, then simplified it to an isotropic and symmetric problem by neglecting air resistance and track friction and assuming that all the horses were identical and spherical."

"Perfect strategy," JJ commented. "Next time he should team up with the engineer and the statistician and refine that model: they'll take the bookies to the cleaners."

"And here is the second story," I said.

On a beautiful summer day, a mathematics professor is visited in his office by a troubled-looking student.

"Sir, I've just seen my grade in your course, and it's an F."

"Yes, I know. You are the only one in the class who failed. You didn't write anything in your answer book."

"But, sir, you don't understand: I *must* pass, otherwise Curly will die."

"Who's Curly? And why will he die if you don't pass?"

"Curly's my pedigree dog. He was abducted on the eve of the final. In the ransom note, the dognappers said that they would kill Curly if I didn't get a pass in your course. My dog is my life, sir. I'll do anything to save him. Please, sir, give me another chance."

The professor, who has heard it all, thinks that this ludicrous and childish appeal must be prompted by enormous desperation, and decides to act humanely. "Yours is the most unusual request for clemency I've ever heard. Normally I'd say that I don't care about your dog and that, if he meant that much to you, you should've prepared for the exam well and passed it along with the other students. But this beautiful weather makes me feel lenient, so I'll give you another chance. It'll have to be a *viva voce*[5] exam, though, since I haven't got the time to prepare a special paper for you. Be in our usual classroom tomorrow at 10 am."

The student thanks the professor and leaves. When he turns up the next morning, he sees a question written on the board.

"Solve this," the professor tells him bluntly.

The student goes to the board, looks at the question, and stands there, fidgeting. "I'm sorry, sir," he explains, "but I worked very hard throughout the night and my mind has gone blank."

"Let's try something else," the professor says, and asks another question.

The student looks at his shoes and says nothing.

"Well? Do you know the answer?"

The only sound in the room is the soft hissing of the air conditioning.

"Hmm..." the professor mumbles, and asks a third question.

No response from the student, who remains motionless and quiet—a statue of total ignorance.

The professor tries one more time, still to no avail, then finally gives up.

"You know nothing at all," he says. "It's a lovely day outside: the sun is shining, the birds are chirping, the trees are in bloom. My thoughts are on vacation and I find myself in a good mood, which you insist on spoiling. All right, let's see if at least you *think* mathematically. How many light bulbs are in this room?"

It is an old-fashioned room, with bulbs hanging symmetrically from the ceiling. The student suspects that the professor might be trying to make fun of him as payback for his miserable performance. Looking at the man, however, he detects no sign of sarcasm so is forced to assume that the question is genuine. Uncertainly, he raises his eyes and counts: 1, 2, 3, ..., 8, then says, "There are eight light bulbs in this room, sir."

The professor shakes his head with sadness. "You don't know the first thing about mathematical thinking. You inspect your little universe and trust in the simplistic version perceived by your senses. You don't go beyond that. In mathematics, young man, you must always be prepared for the unexpected." And the professor puts a hand in his pocket and brings out a bulb. "See what I mean? There are *nine* light bulbs in this room. I cannot—"

---

[5] Oral. (Latin)

"But, sir, my dog—"

"You don't expect me to give you a pass for nothing, do you?" the professor says, exasperated by the student's insistence. "The best I can do is let you off with an incomplete. Come back before the start of the next semester and we'll do another *viva voce*. But *that* will be *that*."

In the fall, the student meets the professor in the same classroom and again is incapable of answering any of his questions. The professor is about to announce his failure, when suddenly he remembers the summer incident.

"You're the one who couldn't even think mathematically, aren't you?"

"Yes, sir."

"Well? Did you learn anything from that experience?"

"Yes, sir, I certainly did."

The professor gives him a long look and says, "I wonder... Tell you what: if you can show me that you think mathematically, I'll give you a pass. How many light bulbs are in this room?"

The student beams, looks up at the ceiling, and counts: 1, 2, 3, ..., 8, then proclaims with confidence, "There are nine light bulbs in here, sir!"

The professor grimaces and shakes his head. "For goodness sake, man! You think I carry a light bulb with me every day?" And he turns both his pockets inside out. "See? No light bulb. The answer to my question is eight. You've learned nothing at all."

"Oh, sir, but I have," the student says calmly, and, reaching into his own pocket, produces a light bulb. "The answer, I think, is still nine."

JJ's eyes glinted. "Depending on whose side you are on—the student's or the professor's—you may stop here or continue by saying that, after seeing the student's bulb, the professor opens his briefcase and produces another, wiping the smugness off the student's face."

"That would be sheer cruelty, and I won't condone it. But back to Gan's lecture: I'm surprised he didn't get a question about his very first assumption, that the universe is infinite. Well? Is it? Have you obtained unimpeachable proof one way or the other?"

"Wouldn't you like to know?" JJ put his hands on the table. "Thanks for the stories," he said. "I've really enjoyed our little chat. If all goes well, we'll talk some more next time." Then, unceremoniously, he sprang to his feet and promptly walked out of the bar.

Notes After the Meeting

Here are the answers to JJ's questions together with some brief personal comments.

Solution to Question 1. The error consists in accepting (24.2) to be valid at $t = T$. Formula (24.2) is derived from (24.1), which is indeed valid at $t = T$; however, at that moment we have

$$x(T) = a, \quad y(T) = 0,$$

and, according to (24.6),
$$y'(T) = -\sqrt{2ga},$$
so (24.1) reduces to
$$ax'(T) = 0,$$
and we deduce that
$$x'(T) = 0.$$
Hence, at $t = T$ equality (24.2) assumes the indeterminate form
$$y'(T) = \frac{0}{0},$$
which gives us no information. The correct value of $y'(T)$ is the one above, as computed in (24.6).

This problem illustrates the need for care and attention required of the members of our scientific community in their handling of mathematical models.

A proper analysis of black holes—much more complex than the simplistic argument used in part (ii) of the 'proof'—would involve the mathematical concept of singularity, which goes beyond the scope of our discussion. All we need to remark here is that, since the assertion of the theorem is false, physicists are safe to pursue their studies.

The reader should note that mathematics is *not* a physical science.

Solution to Question 2. A line has infinite length, whereas a circle is a finite-length curve. Would it, therefore, be correct to suspect that the line has more points than the circle? One thing is clear: they both have infinitely many points. So how do we compare infinities? Can one set be 'more infinite' than another?

A reasonable idea for finding out which of the two sets is the 'larger' one is to pair off their points. Thus, if we cut the circle at one of its points and flatten out its circumference, we could place it on top of the line (L) as a finite length segment AB (see Fig. 24.2). Then we associate each point on this segment with the corresponding (identical) point on the line and notice that, after exhausting all the segment points in this way, we've still got infinitely many points on the line (outside the segment AB) that have no matching pair on the segment. Consequently, we would be hard pressed not to accept that the line has more points than the circle.

Fig. 24.2.

But here is another way of making a pairwise comparison.

We draw the circle tangent to the line (L) at T and denote by O the point on the circle that is diametrically opposite to T (see Fig. 24.3).

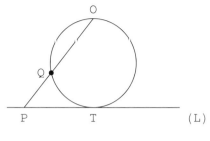

Fig. 24.3.

Picking an arbitrary point P on (L), we join it with O and denote by Q the unique point of intersection of OP with the circle. Clearly, if we change, however slightly, the position of P on (L), we would get a different point Q on the circle. Hence, we can pair off each point on the line (L) with one and only one point on the circle. In particular, point T is paired with itself because it belongs to both the line and the circle. However, the circle has an extra point, O, that has no pair on the line. So it would seem that the circle contains more points than the line.

A superficial examination of these arguments might suggest that they are both equally plausible. But since in mathematics we don't really expect to have it both ways, we must assume that the paradox has an explanation. And it does: the 'size', called cardinality, of an infinite set is defined differently from that of a finite one. It turns out that the line and the circle have the same cardinality.

The intuition we acquire from working with finite sets does not serve us well when we deal with infinite sets. An illustration of this is provided by the famous paradox of the Grand Hotel, conceived by Hilbert.[6] If a hotel with finitely many rooms (which operates a strict policy of one and only one guest per room) is full, then it cannot accommodate any additional guest. This is not the case in a hotel with infinitely many rooms. The manager of the latter can move the guest in room 1 to room 2, the guest in room 2 to room 3, and so on, thus vacating room 1 for the newly arrived person.

Cantor[7] developed a theory of hierarchical 'transfinite' cardinalities and showed, among other things, that the set of real numbers is 'larger' than the set of natural numbers. Since he was not able to evaluate the 'difference' between the cardinalities of these two sets, he formulated the *continuum hypothesis*, according to which there is no set whose 'size' is strictly between that of the integers and that of the real numbers. It turns out that this assertion can

---

[6] See footnote 4 on p. 44.

[7] Georg Ferdinand Ludwig Philipp Cantor: Russian-born German mathematician, 1845–1918.

neither be proved nor disproved within the current framework of set theory.

Infinity, in the shape of a so-called point at infinity, is encountered in perspective artwork, where it is used to create paintings that intend to give a realistic feel for distance in their imagery.

The idea of infinity is often abused by some nonmathematicians. For example, many people construct phrases like 'thesis A has infinitely more supporters than thesis B'. No, it doesn't, for even if every soul on Earth supports thesis A, the supporters will still form a finite set. What these people really mean by 'infinitely more' is 'an awful lot more', or 'far more', but they prefer an exaggeration for emphasis and stylistic purposes, which is fair enough. We all need to make a living.

A Word to the Wise

Infinity

Infinity, denoted by ∞, is a symbol, not a number. It is a powerful but delicate concept, which must be treated, shall we say, with infinite care. One hears it sometimes referred to as being 'larger than any number'. This implies a comparison of infinity with numbers, which is wrong and might make the incautious think of infinity as being itself a kind of number. And yet, for practical purposes, it is useful to design an algebraic formalism for ∞ (and its opposite 'pole' $-\infty$), with the strict proviso that all its 'operations' must be interpreted in a limiting sense, as prescribed in calculus.

In all the limiting processes below we assume that $n \to \infty$.

☺ DO

Symbolically, one can write[8]

$$\infty + \infty = \infty,$$
$$\frac{a}{\infty} = 0 \text{ (for any fixed number } a\text{)}.$$

 DON'T

The following symbolic operations are wrong:

 $\infty - \infty = 0,$

 $\dfrac{\infty}{\infty} = 1,$

 $0 \times \infty = 0.$

---

[8] But not in the presence of a mathematician.

For example,
$$n \to \infty,$$
$$n^2 \to \infty,$$
$$n + a \to \infty,$$
but
$$n^2 - n = n^2\left(1 - \frac{1}{n}\right) \to \infty,$$
$$n - n^2 \to -\infty,$$
$$(n + a) - n \to a,$$
which means that $\infty - \infty$ cannot be properly defined.

The symbolic operation $\dfrac{\infty}{\infty}$ is also impossible to define, as can be seen from the fact that
$$\frac{n^2}{n} = n \to \infty,$$
$$\frac{n}{n^2} = \frac{1}{n} \to 0,$$
$$\frac{an + 1}{n} = a + \frac{1}{n} \to a.$$

Finally, the invalidity of the symbolic operation $0 \times \infty$ can be deduced from the limits
$$\frac{1}{n} \times n^2 = n \to \infty,$$
$$\frac{1}{n^2} \times n = \frac{1}{n} \to 0.$$

☺ DO

Certain other symbolic operations with ∞ are tolerated; thus,
$$\infty^x = \infty \ \text{(for any } x > 0\text{)}, \quad \ln \infty = \infty,$$
$$x^\infty = \begin{cases} 0 & \text{if } 0 < x < 1, \\ \infty & \text{if } x > 1, \end{cases}$$
$$x^{-\infty} = \begin{cases} \infty & \text{if } 0 < x < 1, \\ 0 & \text{if } x > 1. \end{cases}$$

☠ DON'T

A symbolic operation such as
$$☠ \quad 1^\infty = 1$$

is invalid when 1 itself is the result of a limit. This can be seen from the simple examples
$$1^n \to 1,$$
$$\left(1 + \frac{1}{n}\right)^n \to e,$$
where $e \cong 2.718281828459$ is the base of the natural logarithm.

Similar considerations apply to symbolic operations involving $-\infty$.

It should be pointed out once again that, since ∞ is not a number, the best (and correct) way to operate with it is by treating it as a limit.

SCAM 25

The Virtues of Mathematics

> *So I went to work: and here I must needs observe, that as reason is the substance and original of the mathematics, so by stating and squaring everything by reason, and by making the most rational judgment of things, every man may be in time master of every mechanic art.*
> Daniel Defoe[1]

> *Ich behaupte aber, dass in jeder besonderen Naturlehre nur so viel eigentliche Wissenschaft angetroffen werden koenne, als darin Mathematik anzutreffen ist.*[2]
> Immanuel Kant[3]

I met JJ again at a mathematics conference sponsored by Yale University in New Haven, Connecticut. Just like the first time, he was sitting alone at a table in a corner of the hotel bar, gazing at me across the room with dark, inscrutable eyes.

As soon as I sat down, he pushed a piece of paper in my direction. "Here," he said laconically. "For your students."

I glanced at the paper and smiled. The text read something like this.

Question 1 (requires basic geometry). *What, if anything, is wrong with the following assertion and proof?*

Theorem. *Mathematics is divine.*

Proof. *Take a square of cardboard of side 8 cm and cut it along the lines shown in Fig. 25.1 into two equal right trapezoids and two equal right triangles.*

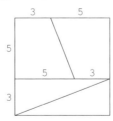

Fig. 25.1.

---

[1] English writer and journalist, c1660–1731. From *Robinson Crusoe*.
[2] However, I claim that in any particular physical science only so much real science can be found as the mathematics applied in it. (German)
[3] German philosopher, 1724–1804.

Now rearrange the trapezoids and triangles into a rectangle, as shown in Fig. 25.2.

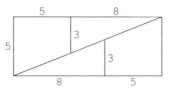

Fig. 25.2.

The area of the original square is $8 \times 8 = 64$ cm^2, whereas the area of the rectangle is $5 \times 13 = 65$ cm^2.

Thus, mathematics has created 1 cm^2 out of nothing. But such an act of creation implies divine intervention. Therefore, mathematics is divine.

Question 2 (requires basic algebra). On his tenth birthday Fred receives a big box of chocolates from his parents. After he eats as many of them as he could without getting sick, he decides to share the remaining 29 pieces among his two brothers, aged 11 and 12, and his three sisters, aged 7, 8, and 9. The siblings start quarreling at once about how the chocolates should be divided. With no agreement in sight and things getting uglier by the minute, Fred steps in and rules that, since the chocolates are his and it is his birthday, he should have the final word. After a little more squabbling, it is finally decided that

(i) each of the two brothers will get an equal share;

(ii) each of the three sisters will also get an equal share;

(iii) a brother-share will be larger that a sister-share, but less than double;

(iv) no share will consist of a single piece.

Fred scratches his head for a bit, then solves the problem. How does he distribute the 29 chocolates?

"Interesting," I said. "Thanks." I folded the paper and put it in my pocket. When JJ did not broach, as usual, a subject of his choice, I prodded him, "What are we going to talk about today?"

"*De omni re mathematica et quibusdam aliis.*[4] How often do you get asked why mathematics is important enough to be studied by all in school and, later, by many in college?"

"Quite frequently," I admitted. "You?"

"All the time. And when I'm not asked, I make a point of saying it myself."

---

[4] Of every thing mathematical and a few others: paraphrase of Pico della Mirandola's famous *De omni re scibili* (Of every knowable thing) with Voltaire's later humorous addition *et quibusdam aliis*. (Latin)

Giovanni Pico della Mirandola: Italian philosopher and scholar, 1463–1494.

Voltaire: the pen name of François Marie Arouet, French writer, essayist, and philosopher, 1694–1778.

"Some people think that our claim concerning the importance of mathematics is nothing more than a sign of professional arrogance."

"Not if you explain it. Gan's university has developed a very entertaining and educationally useful custom. Every other month they organize a mock trial for the general public, in which an academic subject is placed in the dock, so to speak, charged with claiming to be the most important of all subjects. The audience acts as a grand jury, and a faculty member as defense attorney. Last year they invited Gan to go batting for mathematics. It was a very stimulating session, he told me, with members of the audience—call them Aud, generically—asking all manner of questions. The discussion, recorded on Gan's VOTSIT, went something like this.

AUD: I'm a journalist. Why did I have to learn algebra in school when I knew full well that I would never need it in my future career?

GAN: Succinctly put, mathematics consists of numbers, logic, and the power of abstraction. It's true that not everyone has the ability to cope with *higher* mathematics and its demands on the intellect. But everybody should be exposed to a small dose of its *elementary* branches, like algebra, and make an effort to understand their basics. Weaker algebra students may not remember in later life the formula for the square of a binomial, but will be better equipped to construct a logical argument. In your own case, had you tried harder and paid a little more attention in those algebra classes, you wouldn't have skewed the question the wrong way. Instead, you would've said that it was *after* you had discovered your complete lack of affinity for algebra that you decided to go into a nonmathematical profession.

AUD: I'm a historian. I don't see any point in learning mathematics. My full and comfortable life has never been affected by it. It's more important to learn about history, literature, and arts than calculus. After all, most of math can be done by a calculator or computer. I hate math.

GAN (feels a strong whiff of low MICQ in the air): That's quite a statement, which, as it happens, is wrong in every respect. Let me address the issues you raised in a slightly different order.

First, students should learn history, literature, arts, *and* mathematics. If you miss one of these—and a few other—components in your education, you'll be a semidoct.

Second, only a complete ignoramus can say that calculators and computers can do most of mathematics. As inanities go, this is almost on a par with saying that Earth is flat. Computational devices do no more than what you *program* them to do. They are dumb machines, tools that speed up execution. They cannot *design* a way of solving a problem, however simple. By themselves, they are just pretty pieces of junk on a table.

Third, your full and comfortable life has everything to do with math. Do you enjoy your home? It was built by people who had to figure out angles, tensile strength of materials, and other mechanical aspects—all ultimately based on operations with numbers. You like watching your high-definition TV? It was developed by electronic engineers who used physical laws and, implicitly, mathematical computations to produce it. You like driving your car? It was designed by mechanical engineers, whose work relies heavily on mathematical knowledge. Do you appreciate your good health? It is maintained through the advances made by medicine with its huge variety of drugs, created by chemists, diagnostic and treatment machines, built by engineers, and a better understanding of human physiology, studied in detail by applied analysts. Whichever way you turn, you should see that your comfortable life is based in almost every aspect on someone's knowledge of mathematics.

Fourth, and finally, you may have taken a *dislike* to mathematics for some reason or other—for example, because you had a bad teacher, or your colleagues were better than you at it and you felt frustrated and humiliated, or because you give up easily and don't have the stamina to keep at a problem for longer than 30 seconds. It seems to me that you just don't understand what mathematics is about, and you are confusing this with hate for the subject. To hate something simply because you don't understand it is very dangerous. There's been an enormous amount of grief in the world, caused by exactly this kind of attitude. As a historian, you should be a bit more careful how you use your words.

AUD: I'm an accountant. You said that mathematics is about numbers. So is accountancy. What, then, is the difference between you and me?

GAN: I always get my sums right. That apart, numbers are just the atoms in the fabric of our science. They were our point of departure. Sometimes numbers are still incidental to our work, but we have moved way past them in conceptual terms and manipulation.

AUD: I'm a philosopher. Is it true that much of modern mathematics is no more than speculation, an exercise of the mind, producing things of intrinsic beauty but with no practical applications?

GAN: Your statement is correct in regard to the aesthetic pleasure and inner poetry generated by mathematical thought, and completely wrong when it comes to its applicability. Mathematics is being created all the time. Its dedicated and poorly paid servants are constantly amassing a wealth of knowledge that may seem to the uninitiated to be no more than sterile and abstruse intellectual gymnastics. When this painstakingly built knowledge reaches a certain quantitative level, a qualitative jump takes place and a new mathematical concept is introduced. Sooner or later, every such concept turns out to have direct practical application. This may not be immediately obvious because mathematical advances of this nature usually occur well ahead of the time when other scientists adopt them as everyday tools. Two classic examples are readily available: Riemannian[5] geometry, which studies spaces with curvature and torsion and played an important role in the development of the theory of relativity, and algebraic structures—groups, rings, and fields—which are widely used in quantum mechanics. The powerful machinery of analysis, both pure and applied, is always called upon to solve mathematical models that engineers set up but cannot solve themselves. So let's not dismiss mathematics as just an elitist, ivory-tower kind of pursuit.

---

[5] Georg Friedrich Bernhard Riemann: German mathematician, 1826–1866.

AUD: I'm a theologian. Is the use of mathematics confined exclusively to the study of the material world, or can it also be extended to spiritual concepts, such as morals and faith?

GAN: There is an early manuscript[6] by Colin MacLaurin,[7] *De viribus mentium bonipetis*,[8] in which the author attempts to discuss the mathematical modeling of the forces that attract the human mind to unspecified 'good things'. The manuscript refers in its conclusions to virtues, vices, and the intensity of future eternal happiness. This naive approach has no basis in reality, but a day may come when nonphysical manifestations of the mind—which are, after all, generated by very physical brain processes—become machine-readable and thus open to mathematical study.

AUD: I'm a writer. You said that mathematics has a poetic side to it. Do mathematicians write real poetry as well? Or does the predominance of the left hemisphere of their brain over the right one suppress their artistic creativity?

GAN: We can write wordy poems if we feel so inclined.

AUD: Give us an example: recite one of your own productions.

GAN: I don't have any. But I can certainly use my logic and sense of rhythm to make one up.

AUD: Can you also give it mathematical meaning?

GAN: You want a lot, don't you? Hmm... let's see... something short and less demanding... perhaps a limerick?[9] Okay, here it is:

> *A student who goes to the mall*
> *Drives over in no time at all.*
>   *Shops, it seems, make her race*
>   *At an infinite pace,*
> *Since no time is triflingly small.*

AUD: I don't quite understand...

GAN: Imagine the simplest scenario, where the student drives steadily and the road joining her starting point and the mall is a straight line; then

$$\text{speed} = \text{distance}/\text{time}.$$

For a fixed distance, as time decreases and draws closer and closer to zero (becoming triflingly small[10]), the speed increases beyond any bound. In the limiting case of 'no time at all', that is, time $= 0$, the speed is infinite. You cannot divide by zero.

AUD: I'm sorry I asked.

---

[6] MS 3099.15.6 in the Colin Campbell Collection at the Edinburgh University Library. I am grateful to Ian Tweddle for bringing this item to my attention.

[7] Scottish mathematician, 1698–1746.

[8] Concerning the good-seeking forces of minds. (Latin)

[9] A type of amusing five-line verse, named after a city in Ireland.

[10] Not for mathematicians, who are very serious about arbitrarily small quantities and handle them with precision and elegance.

GAN: Maybe you prefer a haiku?[11] Then how about this one:

> *Acknowledge the truth:*
> *That in the land of science*
> *Mathematics rules.*

While these hurried efforts are not going to make their way into any literary anthology, they still show that a mathematician is quite capable of formally using your professional tools. Could you do the same with mine? Could you, for example, wax lyrical about something like homeomorphisms on spaces of continuously differentiable functions?

AUD: I'm a chemist. I took a few courses in mathematics at university and enjoyed them. But I must admit that I have trouble trying to understand the two theories you mentioned earlier: relativity and quantum mechanics. Can you give some simple examples that would help a nonexpert like myself get a glimpse of how powerful mathematics is in underpinning these theories?

GAN: Unfortunately, since you don't have an adequate mathematical background it's hard for me to find such examples, let alone explain them properly. However, I think I can come up with a couple of anecdotal illustrations. Regarding the theory of relativity, you must've heard of the twin paradox. Suppose that one of a pair of twins takes off and roams around the universe at a prodigious speed, close to the speed of light, while the other remains on Earth. On his return, the wanderer is younger than the twin who never left. The resolution of this paradox involves inertial frames of reference, the effects of acceleration, and the Lorentz[12] coefficient. Einstein, though, came up with a more humorous version. When asked about a means of visualizing his theory, he said, "Put your hand on a hot stove for a minute, and it seems like an hour. Sit with a pretty girl for an hour, and it seems like a minute. That's relativity."[13]

Now the other theory. In classical (Newtonian) mechanics, both the position and velocity (or momentum) of a moving object can be assigned simultaneously with full precision. By contrast, in quantum mechanics, according to the uncertainty principle formulated by Heisenberg,[14] the more accurately the position of a particle is determined, the less information is available about its momentum, and vice versa. Since this may not be easily grasped, here is a two-liner to remember it. One day, a traffic policeman stops Heisenberg on the highway and asks him, "Are you aware how fast you were going, sir?" "No," Heisenberg replies, "but I know exactly where I am."

AUD: I'm a librarian. Although I tried very hard in school, I could never get my mind around mathematics. Is it possible that there are people whose brains are

---

[11] A Japanese lyric verse form consisting of exactly 17 syllables, with a distribution of 5, 7, 5 on 3 unrhymed lines. Mathematicians might be tempted to interpret it as an oblique tribute to prime numbers since 3, 5, 7, and 17 are all primes.

[12] Hendrik Antoon Lorentz: Dutch physicist, 1853–1928.

[13] Clearly, Einstein never went through a sexual harassment awareness program.

[14] Werner Karl Heisenberg: German physicist, 1901–1976.

simply not wired for it?

GAN: *All humans* have a built-in ability to cope with mathematics, at least to some extent. As Robert Heinlein[15] wrote, "Those who cannot understand mathematics are not fully human. At best, they are tolerable subhumans who have learned to bathe, wear clothes, and not make messes in the house." If you wish, I can demonstrate that you have this ability. Right here, right now.

AUD (hesitates): How?

GAN: What is $1 + 1$?

AUD (looks around, expecting a catch): Two.

GAN: There you are: you *can* learn basic mathematics. It's just a matter of having a suitable teacher. The moment that marked the start of human civilization was when primitive man realized the truth of $1 + 1 = 2$.

AUD (female, annoyed): Why primitive *man?* Why not primitive woman? Are you saying that she was intellectually inferior?

GAN: Not at all. In fact, primitive woman was already onto negative integers at that time, but did not want to tell her male partner for fear that she might hurt his fragile ego.

AUD (seemingly not satisfied): And was this $1 + 1 = 2$ person black, brown, red, white, or yellow?

GAN: None of the above. He was green. He took every precaution to be environmentally both aware and friendly.

AUD: Are you, mathematicians, always so curt and sarcastic?

GAN: Only when we are asked silly questions.

AUD: I'm a biologist. Does sex ever come into classroom mathematics?

GAN: If you mean, can mathematics be made to look trendy, alluring, and passion-inspiring, then the answer is yes, absolutely. It depends on the skill of the teacher. Otherwise, mathematics is asexual. But I remember one instance when an instructor was accused of using lewd language in a math class. It happened a long time ago at a university in another galaxy. This professor was solving a problem in the two-dimensional geometric plane, which had been divided into two distinct parts: the finite region S_{in} *in*terior to a rectangle ∂S and the infinite region S_{ex} *ex*terior to it. To help the students visualize the partition, he sketched a diagram on the board, which looked like this:

The next day, the dean told him that a female student had complained about his diagram symbols having sexual connotations. He didn't understand, so he

---

[15] American writer of science fiction, 1907–1988.

sketched the diagram for the dean. "But don't you see?" the dean exclaimed with amusement. "Your notation practically implies that *Sex* is not *Sin*, that there is a clear demarcation line between them!"

AUD: Will you tell us why you think that mathematics is the most important discipline?

GAN: Gladly. And I'll do so by allowing some of the best names in the history of science to speak on its behalf. These 'witnesses' are arranged in chronological order by the year of their birth.

Roger Bacon (English scholar, ?1214–1294): The things of this world cannot be made known without a knowledge of mathematics.

Leonardo da Vinci (Italian painter, architect, engineer, mathematician, and philosopher, 1452–1519): No human investigation can be called real science if it cannot be demonstrated mathematically.

Francis Bacon (English philosopher, statesman, and essayist, 1561–1626): For many parts of nature can neither be invented with sufficient subtilty, nor demonstrated with sufficient perspicuity, nor accommodated unto use with sufficient dexterity, without the aid and intervening of the mathematics; of which sort are perspective, music, astronomy, cosmography, architecture, enginery, and divers others.

Galileo Galilei (Italian physicist, astronomer, and philosopher, 1564–1642): The book [of nature] is written in the mathematical language... without whose help it is humanly impossible to comprehend a single word of it, and without which one wanders in vain through a dark labyrinth.

John Locke (English philosopher, 1632–1704): For in all sorts of reasoning every single argument should be managed as a mathematical demonstration.

Daniel Bernoulli (Swiss mathematician, biologist, astronomer, physicist, and oceanographer, 1700–1782): There is no philosophy which is not founded upon knowledge of the phenomena, but to get any profit from this knowledge it is absolutely necessary to be a mathematician.

Charles Babbage (English mathematician, philosopher, mechanical engineer, and computer scientist, 1791–1871): In mathematical science, more than in all others, it happens that truths which are at one period the most abstract, and apparently the most remote from all useful application, become in the next age the bases of profound physical inquiries, and in the succeeding one, perhaps, by proper simplification and reduction to tables, furnish their ready and daily aid to the artist and the sailor.

Henri Poincaré (French mathematician, theoretical physicist, and philosopher of science, 1854–1912): If God speaks to man, he undoubtedly uses the language of mathematics.

James Jeans (English physicist and mathematician, 1877–1946): From the intrinsic evidence of his creation, the Great Architect of the Universe begins to appear as a pure mathematician.

Albert Einstein: In the beginning (if there was such a thing), God created Newton's laws of motion together with the necessary masses and forces. This is all; everything beyond this follows from the development of appropriate mathematics methods by means of deduction.

What more proof do you need to accept that, as Gauss[16] put it, mathematics is the queen of sciences? If civilization were to be represented as a tree, then mathematics is the sap circulating through its vascular system. Because you don't see it, you may not be aware of its presence, or of the vital role it plays in maintaining the health and continuous growth of the tree. You may cut some of the tree's flowers, or leaves, or branches, or even a big limb or two, and the tree will survive. But stop the sap's flow, and the tree dies. The importance of mathematics to the welfare and progress of humanity cannot be overestimated.

I will now finish with a couple of stories. Many believe that a profession is not robust enough and worthy of the public's respect unless its practitioners are capable of poking fun at themselves. If this is true, then, according to the first story, mathematicians have passed the test with flying colors.

A physicist, an accountant, and a doctor arrive at the Pearly Gates, seeking admission.

"Not so fast," the Gatekeeper says. "Orders have come from above that, owing to overcrowding, we are to let in only those who've done something truly special in their life. You, for example," he turns to the physicist, "what particular achievement are you proud of?"

"I developed a unified field theory, which allows all the fundamental forces between elementary particles to be expressed in terms of a single field."

"What??" the Gatekeeper recoils with revulsion. "You tried to quantify and explain Creation? This is heresy!" And he points to the door of the nearby Dark Elevator: "Down you go, you silly man! Now you," he turns to the accountant, "what great feat did you perform in life?"

"I invented the HMOs."

The Gatekeeper looks daggers at him: "Health Maintenance Organizations! You've tried to make people live longer and delay their preordained appointment with the Almighty! You have tampered with His grand design! And you have the gall to seek a place in Heaven? Away with you!"

At this point, the doctor hangs his head and starts walking in the direction of the Dark Elevator.

"Hey, you!" the Gatekeeper shouts after him. "Where do you think you are going?"

"Well, let's not waste time," the doctor says. "I, too, was in the business of healing, so my destination is clear."

"Wait a minute: what branch of medicine did you practice on Earth?"

"Psychiatry."

"Were you any good?"

"Actually, I was quite famous."

---

[16] Karl Friedrich Gauss: German mathematician and scientist, 1777–1855; acknowledged by many as the greatest mathematical mind of all time.

The Gatekeeper smiles and gestures widely. "In that case, please join us, doc, and go straight to the Throne of Eternal Light."

"What for?" the man asks, perplexed.

"Someone with your skills is badly needed there. GAWD thinks He is a mathematics professor."

And here is the second story, which provides you with a glimpse into how comfortable—or not—various professionals are with numbers.

A company has advertised for a senior position, inviting applications from college graduates in any discipline. The only requirement is that the applicant must show a reasonable level of numeracy.

Eventually, a short list is drawn up and interviews are scheduled.

The first candidate to enter the boardroom has a degree in physics. After all the usual questions are asked and answered, the chairman of the interviewing panel says, "How much is 1 plus 1?"

The physicist notices that there is a chalkboard in the room, so she goes to it, writes down one 1 below another 1, draws a line underneath and, adding upwards, says, "1 plus 1 is 2." She writes 2 below the line, then, adding downwards, says, "1 plus 1 is 2. The result is correct, and it is 2."

The panel members are most impressed. They comment to each other that the physicist did not trivialize the problem: she wrote it down, did the sum in time-honored fashion and also double-checked it. "Thank you very much," they say to the candidate. "We'll be in touch."

The next one is a mathematician. Asked about 1 plus 1, he says, "It depends in what base you consider the numbers. If it's base 2, then the result is one-zero, written 10. If the base is a positive integer strictly greater than 2, then the result is 2."

The panel members are in awe. Not only did the candidate give the correct answer, he also covered all possible cases. They hadn't even thought of this number-base angle.

The third contender is a graduate in engineering. When the 1-plus-1 question comes, he immediately puts his hand in his pocket—and freezes. "I'm terribly sorry, but I forgot my calculator at home. Anyway, I wasn't all that hung up on this job," he says, and leaves the room.

The fourth applicant has a degree in philosophy. To the numeracy question, she answers, "This depends on what you have in mind when you say 'one'. Is it simply the interaction of the gestalt of a set consisting of a sole member with a similar set, or is there an intrinsic acquisition of quality involved, which transcends the mere quantitative nature of each of the two sets? For if you put one apple together with another apple on a table, you get two apples; but if you make one river join with another river, then the result is still one, albeit bigger, river..." And she goes on like this until one member of the panel starts snoring, at which point the chairman thanks the candidate and informs her that the result will be announced soon.

Next to face the panel is a statistician who, when asked, says that, with a confidence level of 0.95, 1 plus 1 is 2.00.

Then comes a lawyer. "In view of Smith v. Jones (1976)," he answers, "the result is 2. However, if you want this in writing, then I'll have to add my bill for correspondence and incidentals to it and the total increases to 5,002. Dollars, of course."

The panel members are very relieved to see him go and tell him that, after due process, the result of the selection exercise will be notified to him by telephone and not by signed letter.

Finally, last on the short list, a candidate with a degree in accountancy enters the room. "How much is 1 plus 1?" the weary chairman asks at the end of the interview. The accountant peers into the panel members' eyes, quietly goes to the door, opens it, looks right along the corridor, then left, closes the door gingerly, walks back to the panel's table and, bending forward, whispers in a conspiratorial voice, "How much would you like it to be?"

The accountant got the job.

Two years later, the company was indicted for financial irregularities and was ultimately disbanded.

"Quite a speech," I said. "Did the audience buy Gan's arguments?"

JJ looked pensive. "You never know what the public at large believes when you try to defend mathematics. From what I've seen, I'm not sure the human species is ready to accept it as the language of fundamental truth. I hope that you and your colleagues will be successful in your attempts to make it happen. The collective planetary MICQ, and therefore the future of your civilization, is in your hands. Guard it well and help it flourish. It will take a lot of effort, but it'll be worth it. Doing nothing is not an option."

A whiff of finality exuded from his words. "So I'll see you at the next conference?" I asked.

For the first time in our strange encounters, JJ seemed to hesitate. "I don't know," he said. "There's a rumor that we may be recalled for a change of assignment. But it's just a rumor. We'll have to see, won't we?" Then, unceremoniously, he sprang to his feet and promptly walked out of the bar.

Notes After the Meeting

Here are the answers to JJ's questions together with some brief personal comments.

Solution to Question 1. Let A, B, C, D, E, F, and G be the points indicated in Fig. 25.3.

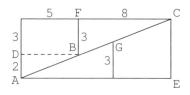

Fig. 25.3.

By Pythagoras's theorem in the right triangle ABD,
$$AB^2 = AD^2 + DB^2 = 2^2 + 5^2 = 4 + 25 = 29,$$
so
$$AB = \sqrt{29} \cong 5.38516 \text{ cm.}$$

The same theorem applied in the right triangle BCF yields
$$BC^2 = BF^2 + FC^2 = 3^2 + 8^2 = 9 + 64 = 73,$$
from which
$$BC = \sqrt{73} \cong 8.54400 \text{ cm.}$$
Then
$$AB + BC \cong 5.38516 + 8.54400 = 13.92916 \text{ cm.} \tag{25.1}$$

If we calculate the length of AC as the hypotenuse in the right triangle ACE, we have
$$AC^2 = AE^2 + EC^2 = 13^2 + 5^2 = 169 + 25 = 194,$$
so
$$AC = \sqrt{194} \cong 13.92839 \text{ cm.} \tag{25.2}$$

From (25.1) and (25.2) we now obtain
$$(AB + BC) - AC \cong 13.92916 - 13.92839 = 0.00077 \text{ cm.}$$

Hence, it is clear that, since $AB + BC > AC$, point B is not on the line AC; in other words, ABC is a 'proper' triangle. By symmetry, so is AGC. The area of the parallelogram $ABCG$ (1 cm^2) has been added to the area of the square to form the area of the rectangle.

Given the exceedingly small difference between $AB + BC$ and AC, triangles ABC and AGC are not visible to the naked eye, particularly when the figures are cut out of cardboard, which adds further inaccuracy to the lines.

This explanation notwithstanding, mathematics is divine.

Solution to Question 2. Let x be a brother-share (as per given condition (i)) and y a sister-share (as per condition (ii)). Then the starting point is, obviously, the equation
$$2x + 3y = 29, \tag{25.3}$$

where x and y are integers satisfying the restrictions

$$y < x < 2y \text{ (condition (iii))}, \quad x, y > 1 \text{ (condition (iv))}. \tag{25.4}$$

On division by 2, (25.3) becomes

$$x + \tfrac{3}{2} y = \tfrac{29}{2}, \tag{25.5}$$

which can also be rewritten as

$$x + y + \tfrac{1}{2} y = 14 + \tfrac{1}{2},$$

or

$$x + y - 14 = -\tfrac{1}{2}(y - 1).$$

Since the left-hand side is an integer, it follows that the right-hand side must also be an integer; in other words, $\tfrac{1}{2}(y-1) = r$, where r is some integer, as yet unknown. This last equality has the equivalent form

$$y = 2r + 1. \tag{25.6}$$

Replacing y in (25.5), we obtain

$$x = \tfrac{1}{2}(29 - 3y) = \tfrac{1}{2}\left[29 - 3(2r+1)\right] = \tfrac{1}{2}(26 - 6r) = 13 - 3r. \tag{25.7}$$

We now need to know what possible values r can take. Using (25.6) and (25.7) in the second restriction (25.4), we see that

$$x = 13 - 3r > 1, \quad y = 2r + 1 > 1,$$

so $r < 4$ and $r > 0$, or, written together, $0 < r < 4$. Since r is an integer, the only values satisfying this double inequality are $r = 1, 2, 3$. The table below shows the corresponding values of x and y.

r	x	y
1	10	3
2	7	5
3	4	7

Of the three pairs of numbers x, y, we see that only one satisfies the first restriction (25.4). This is the middle one, $x = 7$, $y = 5$, which is the unique solution of the problem. (In the pair $(10, 3)$, x exceeds y by more than twice, and in the pair $(4, 7)$, y is larger than x.)

It would be interesting to know how the siblings agreed on restrictions (i)–(iv). The first two were probably accepted because a problem with five different shares would have been badly ill-posed. The third one may have had something to do with the fact that the brothers were older and, therefore, physically much stronger, and that the sisters had not yet been schooled in feminism. As for condition (iv), it is possible that they all considered receiving a single piece to be highly insulting and a particularly fiendish form of torture, for, as is well known, the function of a first piece of chocolate is not to satisfy, but to raise the expectations of the palate for further pieces.

In any case, Fred is a smart young man who cracked the problem by solving a so-called Diophantine[17] equation, which is small work if you want to keep your brothers and sisters happy.

A Word to the Wise

Names of Symbols

Mathematics makes use of a lot of symbols—some familiar, some unusual, and some downright bizarre. Among them, there are a few that are mispronounced by those who don't know their origin or cannot be bothered to utter their full names. Here is a sample of three.

Nabla, written ∇, is a word in Hellenistic Greek meaning a Phoenician[18] harp. The symbol, introduced by Hamilton[19] early in the 19th century for the gradient operator, is called 'del' by nonmathematicians. It seems reasonable to assume that 'del' is an abbreviation of 'delta'. If that's the case, then the 'del' cohorts show not only a lack of mathematical culture, but also of elementary logic: since ∇ is an upside-down Δ (capital delta), the only justifiable alternative to 'nabla' would be 'atled' and not 'del'.

The other two are the hyperbolic sine and hyperbolic cosine, denoted by sinh and cosh, respectively, and defined as functions by

$$\sinh x = \tfrac{1}{2}\left(e^x - e^{-x}\right), \quad \cosh x = \tfrac{1}{2}\left(e^x + e^{-x}\right).$$

While there is no dispute that the spelling of cosh warrants its obvious shortened pronunciation, sinh is variously read as 'shine' or, even more cynically, as 'sinch'.

Not an issue of pronunciation but one of name also arises in connection with the Dirac[20] delta, defined, for example, by

$$\delta(x) = 0 \text{ for } x \neq 0, \quad \int_{-\infty}^{\infty} \delta(x)\,dx = 1.$$

Nonmathematicians using this object, wrongly call it the Dirac delta *function*. This is not a function (if it were, then the integral above would be zero); it is a distribution (generalized function) and should, therefore, be called the Dirac delta distribution or, simply and noncontroversially, the Dirac delta.

---

[17] From the name of the Greek mathematician Diophantus of Alexandria, c214–c284.
[18] Phoenicia: an ancient country at the eastern end of the Mediterranean, on the territory of present-day Syria and Lebanon.
[19] Sir William Rowan Hamilton: Irish mathematician and astronomer, 1805–1865.
[20] Paul Adrien Maurice Dirac: British theoretical physicist, 1902–1984.

Epilogue

> *If we're looking for the source of our troubles, we shouldn't test people for drugs, we should test them for stupidity, ignorance, greed, and love of power.*
>
> Patrick Jake O'Rourke[1]

> *Der Mohr hat seine Schuldigkeit getan, der Mohr kann gehen.*[2]
>
> Friedrich von Schiller[3]

The Yale conference was the last time I met JJ in person. I thought we'd lost contact for good, but a couple of weeks ago, out of the blue, I received an e-mail message from him. It had only three words—"New assignment elsewhere"—and a video attachment, which I immediately played on my computer. Calm and composed as usual, JJ's image spoke to me in the direct, straight-to-the-point manner that I knew so well.

I've read your manuscript and found it to be an accurate account of our conversations. When the book is published, expect all kinds of reactions to it, from full agreement to fierce criticism. But whether my views are judged to be right or wrong isn't important. My objective is simply to make people aware of the issues discussed. Once in possession of the facts, they can form their own opinions, which don't have to coincide with mine. A society that calls itself civilized will not advance unless its members are informed and ready to engage in constructive debate.

As far as I'm concerned, numeracy is by far the best cure for lack of logic. Knowledge changes attitudes and perspectives only when it flows freely into a rational brain. The sooner mankind accepts this truth, the closer it will find itself to the crucial MICQ target of 50.

Naturally, nobody is going to believe that you spoke with an alien. Just for fun, though, when your readers witness disorderly conduct or misbehavior, they should glance at the faces around them and see if they can spot a quiet one observing the scene with dark, inscrutable eyes: that could be Gan, at work with his VOTSIT.

Perhaps some day we'll meet again, to reflect on other pivotal topics we didn't have enough time to cover. Until then, I wish you well—or, as we say in Ganymedean, $**%'$@!

And this is all there is to it. For now.

---

[1] See footnote 1 on p. 183.

[2] The Moor has done his duty, the Moor can go. (German)

[3] See footnote 2 on p. 121. From *Die Verschwörung des Fiesco zu Genua* (translated as Fiesco, or The Genoese Conspiracy).

Index

Academic politics 227
Achilles 99
Algebraic identities 110
    inequalities 215
    operations 76
Andersen, Hans Christian 163
Antisocial behavior 239
Aristotle 217
Asimov, Isaac vii
Average 82, 171, 176, 225

Babbage, Charles 282
Bacon, Francis 26, 282
Bacon, Roger 282
Banach, Stefan 75
    space 75
Baudelaire, Charles 227
Bernoulli, Daniel 282
Big Brother 127
Bijection 89
Birthday paradox 219
Black holes 263
Blackstone, Sir William 203
Brecht, Bertolt 9
BRITE xix, 10
Browne, Sir Thomas 79
Brutus, Marcus Junius 171
Burke, Edmond vii

Caesar, Gaius Julius 173
Calculus, differential 249
    integral 258
    reformed 92
Cantor, Georg Ferdinand Ludwig
      Philipp 271
Capp, Al 157
Cardinality 271
Ceiling 62
Cézanne, Paul 71
Challenge, average age 171
    brand new Porsche 171
    bridge crossing 101

brothers and sisters 56
card-guessing 239
chess and wheat 67
chess tournament 79
chocolate sharing 276
clock face 62
cowboy's horses 33
daughters' ages 146
divine mathematics 275
Earth physics 261
engineering mathematics 251
exit door 34
father-and-son ages 111,
flat Earth 145
four statements 210
Friendly Shell Game 217
GAWD is stronger than
    B.L.Z. Bub 209
global warming 55
graduating students 228
insane population 121
judge and attorneys 195
law is an ass 196
lengths of line and circle 263
locomotive-stopping fly 239
mall and university 218
mathematical fly 43
mind reading 157
nonexistent universe 133
nonprime integer 111
overcharged customers 157
plowing farmers 91
prime generator 251
realtors 183
socket and bulb 227
socks 28
string and cat 185
thickness of folded paper 122
tortoise and hair 91
two envelopes 79
unprovable theorems 102
valueless money 133

water plant 67
wine bottles 19
Chandler, Raymond 111
Churchill, Sir Winston 203
Cicero, Marcus Tullius 209
Civil litigation 209
Coding 70
Complementary event formula 220
Congruence 183
Constitution 199
Continuum hypothesis 271
Count Dracula 222
Credit cards 149
Criminal legal system 195
Curvature 146

Dalí, Salvador 157
Dangling participle 46
Darrow, Clarence 43
Darwin, Charles Robert 126
da Vinci, Leonardo 282
Defoe, Daniel 275
de Gaulle, Charles 67
Demosthenes 45
Descartes, René 65
Differential 258
  operator 166
Differentiation 249
  chain rule of 250
Diophantus of Alexandria 288
Dirac, Paul Adrien Maurice 288
  delta distribution 288
Dirichlet, Johann Peter Gustav
  Lejeune 28, 57
Disraeli, Benjamin 217
Duport, James 121

Einstein, Albert 266, 280, 282
Emerson, Ralph Waldo 9, 91
Equality 30
  approximate 41
  exact 41
Equations, Maxwell's 48
  quadratic 120
  roots of 143
Ethnomathematics 106
Euclid 201
Euler, Leonhard Paul 66

Evolution of knowledge 261
Exponents 170

Factorial 62
Factoring 143
Feminism 104
Fermat, Pierre de 109
  last theorem 109
Fibonacci 164
  sequence 164
Floor 62
Force, centrifugal 125
  gravitational 125
Foreign countries 67
Fractals 164
Fractions 39, 89
  operations with 99
French Revolution 138
Frost, Robert 91
Function, composite 250
  continuous 165, 247
  derivative of 94, 248, 250
  differentiable 249
  exponential 170
  linearly behaved 16
  logarithmic 17
  power 170
  square 17, 110
  square root 16
  trigonometric 208
Functional 141

Galilei, Galileo 282
Galois, Évariste 12
Gauss, Karl Friedrich 283
Geometry 193
  non-Euclidean 201
Giraudoux, Jean 195
Gödel, Kurt 109
Goethe, Johann Wolfgang von 55, 88
Goldbach, Christian 109
  conjecture 109
Golden ratio 164
Gounod, Charles 88
Grammar 43
Grand Hotel paradox 271
Greek alphabet 58

Hamilton, Sir William Rowan 288

Heinlein, Robert 281
Heisenberg, Werner Karl 280
Herbert, Frank 261
Hesiod 239
Highway code 122
Hilbert, David 44, 271
Hoover, Herbert 251

Implication 30
Indefinite integral 237, 258
Induction, engineering 258
  mathematical 178, 255, 258
Infinite sequence 238
Infinity 272
Integration, arbitrary constant of 237, 258, 263
  by parts 251
Interest rates 149

Jeans, James 282
Jefferson, Thomas 192

Kandinsky, Wassili 159
Kant, Immanuel 275
Kennedy, John Fitzgerald 25
Kissinger, Henry 234
Koch, Niels Fabian Helge von 164
  snowflake 164

Lagrange, Joseph-Louis 57
Language 43
  foreign 55
Laplace, Pierre-Simon, Marquis de 57
Leacock, Stephen 111
Lessing, Doris 101
Limiting process 134, 272
Limits 238, 273
Lingua franca 56
Living on debt 145
Locke, John 282
Logarithm 181
Lorentz, Hendrik Antoon 280
Lurie, Alison 67

MacLaurin, Colin 279
Marx, Karl 181

Mathematical commandments 9
  model 4, 266, 267, 270, 278
Maxwell, James Clerk 48
McIlvanney, William 195
Mean, arithmetic 225
  geometric 225
  harmonic 225
Median 82, 226
Mencken, Henry Louis 19
Mephistopheles 88
Meteorologists 218
Metric 141
  space 238
  system 136
MICQ xix, 1, 4
Minkowski, Hermann 57
Mode 226
Modern art 157

Nabla 288
Neumann, John von 52
Newton, Sir Isaac 262
Normed space 141
Number system 54
Numbers, complex 16, 54, 65
  decimal 86
  octal 96
  prime 51, 109, 162

Operation, binary 265
  commutative 265
O'Rourke, Patrick Jake 183, 289
Orwell, George 101, 127, 239
Overqualification 27, 117
Ovidius Naso, Publius 55

Pagnol, Marcel 251
Parallel postulate 201
Pavlov, Ivan Petrovich 45
Percentages 145, 156
  compound 82
Picasso, Pablo Ruiz y 168
Pico della Mirandola, Giovanni 276
Poincaré, Henri 282
Political correctness 101
Polygonal line 134
Polynomial 84
  quadratic 85, 120

Pope, Alexander 25
Probability 217, 225, 226
Problem, ill-posed 86, 89, 287
  well-posed 86
Progression, arithmetic 76
  geometric 53, 76
Proof by contradiction 206
Public media 183
  school system 19
Puccini, Giacomo Antonio Domenico
  Michele Secondo Maria 69
Pulitzer, Joseph 186
Punctuation 43
Pythagoras 79, 237, 261, 286

Quantum mechanics 280

Rectilinear motion 247
Relativity 280
Residue classes 70
Rogers, William Penn Adair 27
Rolle, Michel 57
Rule of precedence 77

Saxon genitive 48
Schelling, Felix 19
Schiller, Friedrich von 121, 289
School mathematical education 33
Schwarz, Karl Hermann Amandus 57
Schweitzer, Albert 261
Semidoct 22, 25, 60, 188, 277
Seneca, Lucius Annaeus 21, 79
Series, absolutely convergent 181
  infinite 53, 171
Shakespeare, William 1, 173, 209

Shanker, Albert 22
Shaw, George Bernard 44
Shell Game 224
Slowism 108
Smith, Adam 199
Socrates 6
Spielberg, Steven 1
Square roots 132
Standard deviation 226
Statistics 217, 225
Stevenson, Robert Louis 171
Stoker, Bram 222
Swift, Jonathan 227

Thatcher, Margaret 105
Trigonometry 207
Truesdell III, Clifford Ambrose 48
TV advertising 111
Twain, Mark ix

Units of measurement 133

Vergilius Maro, Publius 68
Vlad III, Prince of Wallachia 222
Voltaire 276
VOTSIT xix, 6

Weather forecasting 222
Wilde, Oscar 171
Wittgenstein, Ludwig 43

Yeats, William Butler 183

Zeno of Elea 99